Photon Counting Computed Tomography

Scott Hsieh • Krzysztof (Kris) Iniewski
Editors

Photon Counting Computed Tomography

Clinical Applications, Image Reconstruction and Material Discrimination

 Springer

Editors
Scott Hsieh
Mayo Clinic
Rochester, MN, USA

Krzysztof (Kris) Iniewski
Emerging Technologies CMOS Inc.
Port Moody, BC, Canada

ISBN 978-3-031-26064-3   ISBN 978-3-031-26062-9   (eBook)
https://doi.org/10.1007/978-3-031-26062-9

This Springer imprint is published by the registered company Springer Nature Switzerland AG
The registered company address is: Gewerbestrasse 11, 6330 Cham, Switzerland

Contents

Editors Biography

Scott Hsieh is an assistant professor at the Mayo Clinic whose research includes modeling and development of photon counting detectors. His past contributions to photon counting include theory for circuit designs that could reduce dose and improve contrast, and simulations to define application requirements. Previously, he was a faculty member at UCLA and an instructor at Stanford. He is a named inventor on 13 issued US utility patents.

Krzysztof (Kris) Iniewski is a director of detector architecture and applications at Redlen Technologies Inc. in British Columbia, Canada. During his 16 years at Redlen, he has managed development of highly integrated CdZnTe detector products in medical imaging and security applications. Prior to Redlen, Kris hold various management and academic positions at PMC-Sierra, University of Alberta, SFU, UBC, and University of Toronto.

Dr. Iniewski has published over 150+ research papers in international journals and conferences. He holds 25+ international patents granted in USA, Canada, France, Germany, and Japan. He wrote and edited 75+ books for Wiley, Cambridge University Press, McGraw Hill, CRC Press, and Springer. He is a frequent invited speaker and has consulted for multiple organizations internationally.

Part I
Clinical Applications

Medical Photon-Counting CT: Status and Clinical Applications Review

Thomas Flohr, Martin Petersilka, Andre Henning, Stefan Ulzheimer, and Bernhard Schmidt

1 Introduction

Computed tomography (CT) is the backbone of radiological diagnosis. The application spectrum of CT has been continuously expanded by technological advances, among them the introduction of spiral CT [1], the rapid progress of multi-detector row CT [2, 3], and new system concepts such as wide detector CT [4] or dual-source CT [5] with their specific clinical benefits.

Today, CT is a mature modality in its saturation phase. Yet, there are still clinical limitations for current CT technology, mainly caused by insufficient spatial resolution for some applications such as CT angiography (CTA) of the coronary arteries and limited potential to further reduce the radiation dose to the patient. Furthermore, CT is in general very sensitive, but not very specific – the morphological information obtained with a CT scan is often not sufficient to guide clinical decisions without further work-up of the patient. Dual-energy CT has gained momentum as a technique to enhance the clinical value of CT by adding functional information to morphology. Reviews on clinical applications of dual-energy CT may be found in [6–12]. Dual-energy CT data can currently be acquired with dual-source CT systems [5, 13], CT systems with fast kV switching [14], or dual-layer detector CT systems [15]. However, each of these solutions has inherent limitations.

Photon-counting detectors are a new technology with the potential to overcome these drawbacks, by providing CT data at very high spatial resolution, without electronic noise and with inherent spectral information. Photon-counting detectors and their potential benefits have already been evaluated in experimental CT benchtop systems more than 10 years ago (e.g., [16]). The detectors used in these

T. Flohr (✉) · M. Petersilka · A. Henning · S. Ulzheimer · B. Schmidt
Siemens Healthcare GmbH, Computed Tomography, Forchheim, Germany
e-mail: thomas.flohr@siemens-healthineers.com

early systems, however, did not tolerate the high x-ray flux rates required in medical CT, and stable image quality without ring artifacts caused by signal drift and other processes could not be achieved. Significant progress in detector material synthesis and detector electronics design has meanwhile helped to overcome these problems, and pre-clinical whole-body photon-counting CT prototypes could be installed for pre-clinical testing in human subjects. Meanwhile, a commercial CT-system with photon-counting detectors has been released.

This review article gives an overview of the basic principles of medical photon-counting detector CT and of the clinical experience gained so far in pre-clinical installations. Other reviews of photon-counting detector CT are available in [17–20].

2 Principles of Medical Photon-Counting CT

2.1 Properties of Solid-State Scintillation Detectors

Solid-state scintillation detectors are used in all current medical CT scanners. They consist of individual detector cells with a side length of 0.8–1 mm, made of a scintillator with a photodiode attached to its backside (see Fig. 1). The absorbed x-rays produce visible light in the scintillator which is detected by the photodiode and converted into an electrical current. Both the intensity of the scintillation light

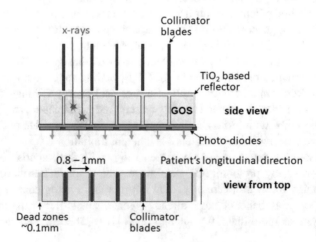

Fig. 1 Schematic drawing of an energy-integrating scintillator detector. Side view and view from top. Individual detector elements made of a scintillator such as gadolinium oxide or gadolinium oxysulfide (GOS) absorb the x-rays (red arrows) and convert their energy into visible light (orange circles). The light is registered by photodiodes and converted into an electrical current. The detector elements are separated by optically in-transparent layers (e.g., based on TiO$_2$) to prevent optical crosstalk – these layers are "dead zones." Collimator blades above the separation layers suppress scattered radiation

and the amplitude of the induced current pulse are proportional to the energy E of the absorbed x-ray photon. All current pulses registered during the measurement time of one reading (projection) are integrated. X-ray photons with lower energy E, which carry most of the low-contrast information, contribute less to the integrated detector signal than x-ray photons with higher energy. This energy weighting reduces the contrast-to-noise ratio (CNR) in the CT images. This is a particular challenge in CT scans after administration of iodinated contrast agent – the majority of all medical CT scans. The x-ray absorption of iodine is highest at lower energies closely above its K-edge at 33 keV; these energies are down-weighted in the signal of a scintillation detector.

The low-level analogue electric signal of the photodiodes is distorted by electronic noise which becomes larger than the quantum noise (Poisson noise) of the x-ray photons at low x-ray flux. It causes a disproportional increase of image noise and instability of low CT numbers (e.g., in CT scans of the lungs) if the flux is further reduced and sets a limit to radiation dose reduction in medical CT.

The individual detectors are separated by optically in-transparent layers with a width of about 0.1 mm to prevent optical crosstalk. X-ray photons absorbed in the separation layers do not contribute to the measured signal even though they have passed through the patient – they are wasted radiation dose. Current medical CT detectors with a size of about 0.8×0.8 mm^2 to 1×1 mm^2 have a geometric dose efficiency of about 80–90%. Significantly reducing the size of the scintillators to increase spatial resolution beyond today's performance levels while keeping the width of the separation layers constant is problematic because of reduced geometric efficiency.

2.2 Properties of Photon-Counting Detectors

Photon-counting detectors are made of semiconductors such as cadmium-telluride (CdTe) or cadmium-zinc-telluride (CZT). High voltage (800–1000 V) is applied between the cathode on top of the detector and pixelated anode electrodes at the bottom (see Fig. 2, left). The absorbed x-rays produce electron-hole pairs which are separated in the strong electric field. The electrons drift to the anodes and induce short current pulses (10^{-9} s). A pulse-shaping circuit transforms them to voltage pulses with a full width at half maximum (FWHM) of 10–15 nanoseconds; their pulse-height is proportional to the energy E of the absorbed x-ray photons. The pulses are counted if they exceed a threshold T_0 (see Fig. 2, right). In a photon-counting detector for medical CT, T_0 is about 25 keV.

Photon-counting detectors have several advantages compared to solid-state scintillation detectors. The individual detectors are defined by the strong electric field between common cathode and pixelated anodes (Fig. 2, left), and there are no additional separation layers. The geometrical dose efficiency of a photon-counting detector is only reduced by anti-scatter collimator blades or grids. Each "macro pixel" confined by collimator blades may be divided into smaller sub-pixels which

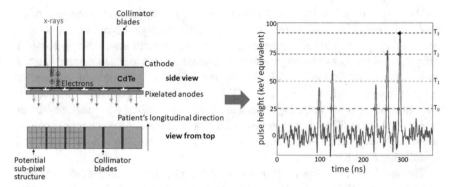

Fig. 2 Left: Schematic drawing of a direct converting photon-counting detector. Side view and view from top. The x-rays (red arrows) absorbed in a semiconductor such as CdTe or CZT produce electron-hole pairs that are separated in a strong electric field between cathode and pixelated anodes and induce fast signal pulses at the anodes. The individual detectors are formed by the pixelated anodes and the electric field; there are no separation layers between them. Collimator blades are needed to suppress scattered radiation. Right: The pulses are counted as soon as they exceed a threshold T_0 (dashed blue line, "counting" is indicated by a blue dot). T_0 has a typical energy of 25 keV. Three additional thresholds at higher energies (T_1 at 50 keV, T_2 at 75 keV, T_3 at 90 keV) are also indicated

are read out separately to increase spatial resolution (indicated in Fig. 2, left. The pixelated anodes must then be structured correspondingly, which is not shown in order to not overload the drawing).

Low-amplitude baseline noise is well below the threshold T_0 (see Fig. 2, right) and does not trigger counts – even at low x-ray flux, only the statistical Poisson noise of the x-rays is present in the signal. CT scans at very low radiation dose or CT scans of obese patients show therefore less image noise, less streak artifacts, and more stable CT numbers than scans with a scintillation detector, and radiation dose reduction beyond today's limits seems possible.

The detector responsivity in the x-ray energy range from 30 to 100 keV is approximately constant. All x-ray photons registered during the time of a projection (reading) contribute equally to the measurement signal regardless of their energy E, as soon as E exceeds T_0. There is no down-weighting of lower energy x-rays as in solid-state scintillation detectors. Photon-counting detectors can therefore provide CT images with potentially improved CNR, in particular in CT scans after administration of iodinated contrast agent.

Several counters operating at different threshold energies can be introduced for energy discrimination (see Fig. 2, right). Physically, the different thresholds are realized by different voltages which are fed into pulse-height comparator circuits. In the example shown in Fig. 2, four different energy thresholds T_0, T_1, T_2, and T_3 are realized. The photon-counting detector simultaneously provides four signals S_0, S_1, S_2, and S_3 with different lower-energy thresholds T_0, T_1, T_2, and T_3. Subtracting the detector signals with adjacent lower-energy thresholds results in "energy bin" data. Energy bin $b_0 = S_1 - S_0$ as an example contains all x-ray photons detected in the energy range between T_1 and T_0.

Simultaneous read-out of CT data in different energy bins opens the potential of spectrally resolved measurements and material differentiation in any CT scan. Established dual-energy applications – mainly based on decomposition into two base materials such as iodine and water, or iodine and calcium – are routinely feasible, and virtual monoenergetic images (VMIs), iodine maps, and virtual non-contrast images (with the iodine removed) or virtual non-calcium images can be computed whenever needed for the diagnosis. Data acquisition with more than two thresholds enables multi-material decomposition under certain preconditions. Unfortunately, the availability of N energy thresholds does not imply potential differentiation of N base materials. Compton scattering and the photoelectric effect are the only two relevant interaction mechanisms in the x-ray energy range accessible to CT (35–140 keV). Their energy dependence is similar for all elements without K-edge in this range – this applies to all materials naturally occurring in the human body including iodinated contrast agent. As soon as two base materials, e.g., water and iodine, have been chosen, the energy-dependent attenuation of any other material can be described by a linear combination of the two base materials. It is therefore not possible to differentiate this other material from a mixture of the two base materials. Differentiation of two base materials requires two measurements at different energies – dividing the energy range into more than two energy bins will not provide relevant new information. The situation changes if a material with K-edge in the energy range accessible to CT, such as gadolinium with a K-edge at 50 keV, is added to the two base materials. For a K-edge material, the energy dependence of x-ray attenuation is different, and CT measurements at three or more energies can be used for three-material decomposition (two base materials plus the K-edge material). Three- or more-material decomposition with CT data in three or more energy bins will be limited to clinical scenarios in which two contrast agents (e.g., iodine and gadolinium, or iodine and bismuth) are applied and need to be separated, or other heavy elements are introduced into the human body (e.g., tungsten or gold nanoparticles).

In addition to potential material decomposition, the CNR of the images can be further improved by optimized weighting of the different energy bins. Instead of just adding the bin data for the reconstruction of an image, higher weights can be assigned to the low-energy bin data, in particular in CT scans using iodinated contrast agent.

2.3 Challenges for Photon-Counting Detectors in Medical CT

Compared to established dual-energy CT acquisition techniques, photon-counting detectors are often assumed to provide better energy separation and less spectral overlap. However, unavoidable physical effects reduce the energy separation of CdTe- or CZT-based photon-counting detectors. The current pulses produced by x-rays absorbed close to pixel borders are split between adjacent detectors ("charge sharing"). This leads to registration of a high-energy x-ray photon as several lower-energy events. Cd and Te have K-edges at 26.7 and 31.8 keV, respectively. Absorbed x-rays kick out K-electrons of the detector material and lose the K-edge energy. The empty K-shells are immediately refilled, and characteristic x-rays are released which are counted in the same detector pixel or "escape" to neighboring pixels and are counted there ("K-escape"). In summary, high-energy x-ray photons are erroneously counted at lower energies, and spectral separation as well as spatial resolution (in case of K-escape) is reduced. Charge sharing and K-escape are illustrated in Fig. 3, top. If the detector is read out in several energy bins, the low-energy bins will contain wrong high-energy information ("high-energy tails"). Increasing the size of the detector pixels reduces the high-energy tails, because boundary effects such as charge sharing and K-escape contribute less to the total detector signal (see Fig. 3, bottom). For a realistic detector model with detector sizes as in Fig. 3, including

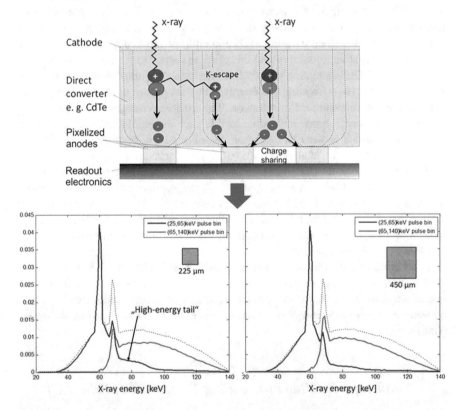

Fig. 3 Top: Schematic illustration of charge sharing at pixel boundaries and K-escape, which result in multiple counting of x-ray photons at wrong lower energies. Bottom: Charge sharing and K-escape lead to a characteristic high-energy tail of low-energy bins. Computer simulation of the x-ray spectra recorded in the two energy bins of a realistic photon-counting detector for an incident 140 kV spectrum (bin$_1$, 25–65 keV; blue line; bin$_2$, 65–140 keV, green line). Left: pixel size 0.225 × 0.225 mm^2. Right: pixel size 0.45 × 0.45 mm^2. Increasing the pixel size reduces the high-energy tail

charge sharing, fluorescence, and K-escape, the spectral separation with two energy bins is equivalent to that of a dual-kV technique with optimized pre-filtration [21].

Even larger detector pixels would further improve spectral separation. Unfortunately, there is an effect that limits the maximum size of the detector pixels: the finite width of the voltage pulses after pulse-shaping (FWHM ≥ 10 ns). Medical CTs are operated at high x-ray flux rates up to 10^9 counts per s and mm^2 – if the detector pixels are too large, too many x-ray photons hit them too closely in time to be registered separately. Overlapping pulses are then counted as one hit only ("pulse pileup"). Pulse pileup leads to nonlinear detector count rates and finally to detector saturation. Even though the signal can be linearized before the onset of saturation, significant quantum losses, increased image noise, and reduced energy discrimination cannot be avoided. Finding the optimum size of the detector cells to balance pulse pileup, charge sharing, and K-escape is one of the most challenging tasks in designing a photon-counting detector.

Yet another challenge for photon-counting detectors is count-rate drift at higher x-ray flux rates. Non-homogeneously distributed crystal defects in the sensor material cause trapping of electrons and holes – the resulting space charges modify the electric field distribution differently in the individual detector pixels. Depending on the "irradiation history" of the pixels, the characteristics of the signal pulses are changed, and this deviation from calibration may cause severe ring artifacts in the images. While count-rate drift was one of the major problems for the use of photon-counting detectors in medical CT, it could meanwhile be reduced to clinically tolerable values by refined material synthesis.

3   Pre-clinical Evaluation of Photon-Counting CT

Photon-counting detectors are a promising new technology for future medical CT systems. Currently, pre-clinical prototypes are used to evaluate the potential and limitations of photon-counting CT in clinical practice. We will focus on these pre-clinical installations and leave out other more experimental solutions, benchtop systems, and photon-counting micro CT systems.

A pre-clinical single-source CT system with photon-counting detector based on CZT (Philips Healthcare, Haifa, Israel) provides an in-plane field of view of 168 mm and a z-coverage of 2.5 mm, with a rotation time of 1 s [22]. The size of the detector pixels is 0.5×0.5 mm^2. The photon-counting detector has five energy thresholds. The system has so far been used for scans of phantoms and animals. Improved assessment of lung structures due to higher resolution [22] was demonstrated as well as improved visualization of the in-stent lumen and in-stent restenosis in coronary stents [23]. Differentiation between blood and iodine in a bovine brain was shown by computing iodine maps and virtual non-contrast images [24]. The separation of multiple contrast agents by means of multi-material decomposition and K-edge imaging was evaluated in several studies [25–30], with various potential clinical applications. Separation of iodine, gadolinium, and calcium might be helpful in

CTAs of the aorta for the assessment of aortic endoleak dynamics and distinction from intra-aneurysmatic calcifications in a single scan [27]. Virtual CT colonoscopy might benefit from differentiation of gadolinium-tagged polyps and iodine-tagged fecal material [28]. Sequential administration of several contrast agents might enable imaging of multiple uptake phases in an organ with a single scan [29, 30].

A pre-clinical hybrid dual-source CT scanner is equipped with a conventional scintillation detector and a CdTe photon-counting detector (Siemens Healthcare GmbH, Forchheim, Germany). The photon-counting detector consists of sub-pixels with a size of 0.225×0.225 mm^2. The detector provides two energy thresholds per sub-pixel. 2×2 sub-pixels can be binned to a "sharp" pixel or "UHR" pixel with a pixel size of 0.45×0.45 mm^2, and 4×4 sub-pixels can be binned to a "macro" pixel with a size of 0.9×0.9 mm^2 comparable to today's medical CT systems. By assigning alternating low-energy and high-energy thresholds to adjacent detector sub-pixels in a "chess pattern mode," the detector provides four energy thresholds in "macro" pixels. The in-plane field of view of the photon-counting detector is 275 mm, and the z-coverage is 8–16 mm, depending on the read-out mode. A completion scan with the energy-integrating sub-system can be used to extend the photon-counting field of view to 500 mm. The shortest rotation time of the system is 0.5 s. The x-ray tubes can be operated at voltages up to 140 kV, with a tube current up to 550 mA. The pre-clinical hybrid dual-source CT scanner was described and evaluated in [31–33].

Yet another pre-clinical single-source CT scanner (Siemens Healthcare GmbH, Forchheim, Germany) is equipped with a CdTe photon-counting detector with a similar pixel geometry to the hybrid dual-source CT. Its field of view is 500 mm, and its z-coverage is 57.6 mm (144×0.4 mm) in standard resolution mode ("macro" pixels) and 24 mm (120×0.2 mm) in "UHR" mode. The shortest rotation time of the system is 0.3 s.

The imaging performance of the pre-clinical hybrid dual-source CT was evaluated by means of phantom and cadaver scans [34, 35], demonstrating clinical image quality at clinically realistic levels of x-ray photon flux with negligible effect of pulse pileup [34]. In contrast-enhanced abdominal scans of human volunteers, photon-counting detector images showed similar qualitative and quantitative scores to conventional CT images for image quality, image noise, and artifacts while additionally providing spectral information for material decomposition [36].

The predicted CNR improvements with photon-counting detectors, expected as a result of the missing down-weighting of low-energy x-ray photons, were confirmed both for contrast-enhanced scans (CNR of iodinated contrast agent vs. soft tissue) and non-contrast scans (CNR of different tissues). Improved iodine CNR by photon-counting CT was demonstrated in a study using four anthropomorphic phantoms simulating four patient sizes [35]. A mean increase in iodine CNR of 11%, 23%, 31%, and 38% relative to the scintillation detector system at 80, 100, 120, and 140 kV, respectively, was shown. A similar overall improvement of the iodine CNR by about 25% was already found in [34]. The improvements in iodine CNR can potentially be translated into reduced radiation dose, or reduced amount of contrast agent. Improvement of soft-tissue contrasts was demonstrated in a brain CT study with 21 human volunteers [37]. The higher reader scores for the differentiation of gray and white brain matter for photon-counting CT images compared to conventional CT images were attributed to both higher soft-tissue contrasts (10.3 ± 1.9 HU vs. 8.9 ± 1.8 HU) and lower image noise for photon-counting CT.

The impact of missing electronic noise on image quality was assessed for various clinical applications at low radiation dose. Less streaking artifacts and more homogeneous image noise in shoulder images acquired with the photon-counting detector of the pre-clinical hybrid dual-source CT as compared to its scintillation detector were demonstrated [38]. Symons et al. [39] evaluated the performance of the photon-counting CT prototype for potential low-dose lung cancer screening. Scanning a lung phantom at low radiation dose, the authors found better Hounsfield unit stability for lung, ground-glass, and emphysema-equivalent foams in combination with better reproducibility. Stability of Hounsfield units is an important prerequisite for quantitative CT. Additionally, photon-counting CT showed less noise and higher CNR. The better performance of photon-counting CT at very low radiation dose, attributed to the effective elimination of electronic noise and better weighting of low-energy x-ray photons, might enable further reduced radiation dose in CT lung cancer screening. In a study with 30 human subjects undergoing dose-reduced chest CT imaging [40], photon-counting CT demonstrated higher diagnostic quality with significantly better image quality scores for lungs, soft tissue, and bone, and fewer beam-hardening artifacts, lower image noise, and higher CNR for lung nodule detection (see Fig. 4a, b).

Improved quality of coronary artery calcium (CAC) scoring at low radiation dose was shown in a combined phantom, ex vivo and in vivo study [41]. Agreement between standard-dose (average $CTDI_{vol} = 5.4$ mGy) and low-dose (average $CTDI_{vol} = 1.6$ mGy) CAC score in ten volunteers was significantly better for photon-counting CT than for conventional CT, with better low-dose CAC score reproducibility. This finding was attributed to the absence of electronic noise in combination with improved calcium-soft-tissue contrasts due to missing down-weighting of low-energy x-ray photons. The authors concluded that photon-counting CT technology may play a role in further reducing the radiation dose of CAC scoring.

Improvements in spatial resolution with the pre-clinical hybrid dual-source CT enabled by the smaller pixels of its photon-counting detector in "sharp" mode and in "UHR" mode were evaluated in several phantom and patient studies. 150 μm in-plane spatial resolution, with a cutoff spatial frequency of the modulation transfer function (MTF) at 32.4 lp/cm, and minimum slice widths down to 0.41 mm were demonstrated, and better spatial resolution was confirmed in clinical images of the lungs, shoulder, and temporal bone [42] (see Fig. 5a, b). At matched in-plane spatial resolution, photon-counting images had less image noise than conventional CT images because of the better modulation transfer function (MTF) of the measurement system. Significant improvements of coronary stent lumen visibility in "sharp" and "UHR" mode were found [43], as well as superior qualitative and quantitative image characteristics for coronary stent imaging when using a dedicated

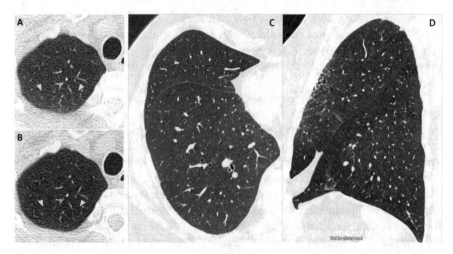

Fig. 4 Lung imaging with photon-counting CT. (**A, B**) Low-dose imaging. Low-dose lung scan acquired with a pre-clinical hybrid dual-source CT prototype with photon-counting detector. (**A**) Scintillation detector image. (**B**) Photon-counting detector image, demonstrating less image noise (arrowheads) because of the absence of electronic noise. Courtesy of R Symons, NIH, Bethesda, USA. (**C, D**) High-resolution imaging. Lung scan of a 74-year-old woman with breast cancer and signs of fibrosis after radiation therapy, acquired with a pre-clinical single source CT prototype with photon-counting detector. Data acquisition: "UHR mode", 120 × 0.2 mm collimation, 0.3 s rotation time, $CTDI_{vol}$ = 3.89 mGy, DLP = 126 mGycm. Image reconstruction: sharp convolution kernel, 1024 × 1024 image matrix, 0.4 mm slice width. Excellent visualization of fibrosis and fine details such as fissures. (Courtesy of Dr. J. Ferda, Pilsen, Czech Republic)

sharp convolution kernel [44]. Figure 5d, e, shows a coronary stent scanned with the photon-counting CT prototype in the "macro" and in the "sharp" acquisition mode.

In a small study with eight human volunteers undergoing scans of the brain, the thorax, and the left kidney, Pourmorteza et al. [45] observed improved spatial resolution and less image noise with the "UHR" mode compared with standard-resolution photon-counting CT in "macro" mode. Substantially better delineation of temporal bone anatomy scanned with the "UHR" mode compared with the ultrahigh-resolution mode of a commercial energy-integrating-detector CT scanner was demonstrated [46].

Lung imaging is another clinical application that might benefit from increased spatial resolution. Superior visualization of higher-order bronchi and third-/fourth-order bronchial walls at preserved lung nodule conspicuity compared with clinical reference images could be demonstrated in 22 adult patients referred for clinically indicated high-resolution chest CT [47]. The authors combined photon-counting CT in "sharp" acquisition mode with image reconstruction at 1024 × 1024 matrix size using a dedicated sharp convolution kernel. According to the authors, photon-counting CT is beneficial for high-resolution imaging of airway diseases, and potentially for other pathologies, such as fibrosis, honeycombing, and emphysema. An example of high-resolution photon-counting chest CT is shown in Fig. 4c, d.

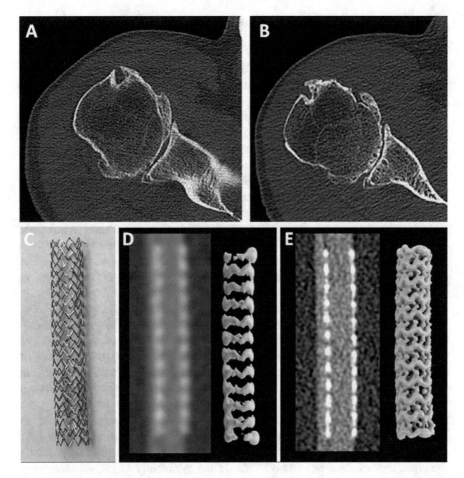

Fig. 5 High-resolution imaging with photon-counting CT. All images are acquired with the pre-clinical hybrid dual-source CT prototype. (**A, B**) Example of a shoulder scan. (**A**) Scintillation detector image. (**B**) Photon-counting detector image in "sharp" mode, demonstrating higher spatial resolution and significantly improved visualization of bony structures. (**C–E**) Coronary stent (**C**) scanned in "macro" mode (**D**), corresponding to the resolution level of today's medical CT systems, and in "sharp" mode (**E**). (Courtesy of Clinical Innovation Center, Mayo Clinic Rochester, MN, USA)

Increased resolution comes at the expense of increased image noise if the radiation dose is kept constant. Increased radiation dose to the patient to compensate for the higher noise may not be acceptable and may not even be available in all clinical situations. Spatial resolution does not only depend on the detector pixel size but also on the focal spot size of the x-ray tube which needs to be correspondingly small. The smaller the focal spot is, the less x-ray tube power is usually available, which may limit the clinical applicability of high-resolution photon-counting CT. Nonlinear data and image denoising techniques will therefore play a key role in harnessing the high-resolution potential of photon-counting detectors (see, e.g., [48–50]).

A key benefit of photon-counting CT is spectrally resolved data acquisition in any scan. Spectral information can readily be added to the anatomical images for better visualization of structures by means of virtual monoenergetic images (VMIs), for material classification and quantitation, and to obtain quantitative information about local perfusion by means of iodine maps and virtual non-contrast images.

The spectral performance of the pre-clinical hybrid dual-source prototype with photon-counting detector was evaluated in phantom studies [51], and the CT number accuracy in VMIs and iodine quantification accuracy were found to be comparable to dual-source dual-energy CT. According to the authors, photon-counting CT offers additional advantages, such as perfect temporal and spatial alignment to avoid motion artifacts, high spatial resolution, and improved CNR. In an anthropomorphic head phantom containing tubes filled with aqueous solutions of iodine (0.1–50 mg/ml) excellent agreement between actual iodine concentrations and iodine concentrations measured in the iodine maps was observed [52]. The authors assessed the use of iodine maps and VMIs in head and neck CTA in 16 asymptomatic volunteers and proposed VMIs at different keV as a method to enhance plaque detection and characterization as well as grading of stenosis.

The routine availability of VMIs with photon-counting CT may pave the way to further standardization of CT protocols. VMIs are based on a two-material decomposition into water and iodine; they show the correct attenuation values of iodine at the selected keV. VMIs at standardized keV levels tailored to the clinical question (e.g., 50–70 keV for contrast-enhanced examinations of parenchymal organs, 40–50 keV for CTAs) may serve as primary output of any CT scan regardless of the acquisition protocol (see Fig. 6). Follow-up scans will then be easily comparable, because their image impression and quantitative CT numbers depend on the standardized keV level only and no longer on the acquisition protocol. Prerequisite for the routine use of VMIs is the availability of refined processing techniques (see, e.g., [53]) to enhance CNR and image quality. Going one step further, the acquisition protocol may be standardized as well. Some authors [54] already recommend a standardized acquisition protocol with 140 kV x-ray tube voltage for contrast-enhanced abdominal CT examinations in all patient sizes, with standardized VMI reconstruction at 50 keV. According to the authors, optimal or near-optimal iodine CNR for all patient sizes is obtained with this protocol.

Several authors assessed the performance of spectral photon-counting CT for detection and characterization of kidney stones, another established dual-energy CT application [55–57]. They found comparable overall performance to state-of-the-art dual-energy CT in differentiating stone composition, while photon-counting CT was better able to help characterize small renal stones [57]. Because of its higher spatial resolution, photon-counting CT can provide both a high-resolution image of the stone structure and a material-map image of the stone composition (see Fig. 7).

If the photon-counting detector is operated with more than two energy bins, multi-material decomposition is possible if K-edge elements are present. In a canine model of myocardial infarction, Symons et al. [58] determined the feasibility of

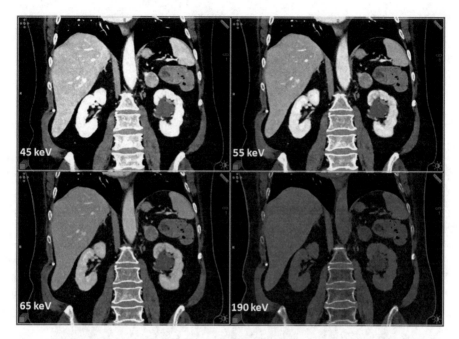

Fig. 6 Use of VMIs as standardized output of photon-counting CT. Abdominal images of a 67-year-old woman with adrenal adenoma and parapelvic renal cyst, acquired with a pre-clinical single-source CT prototype with photon-counting detector. Data acquisition: "UHR" mode, 120 × 0.2 mm collimation, 0.3 s rotation time, $CTDI_{vol} = 10.2$ mGy, DLP = 450 mGycm. Image reconstruction: VMIs at various keV levels, 0.4 mm slice width. Note the decreasing contrast of iodine and calcium with increasing keV. The 45 keV images may serve as standard output of CTAs, the 55 keV or 65 keV images are optimal for examinations of parenchymal organs, and the 190 keV images may substitute virtual non-contrast images with the iodine removed. (Courtesy of Dr. J. Ferda, Pilsen, Czech Republic)

dual-contrast agent imaging of the heart to simultaneously assess both first-pass and late enhancement of the myocardium with the pre-clinical hybrid dual-source CT prototype. In a canine model of myocardial infarction, gadolinium was injected 10 min prior to CT, while iodinated contrast agent was given immediately before the CT scan. The authors concluded that combined first-pass iodine and late gadolinium maps allowed quantitative separation of blood pool, infarct scar, and remote myocardium. The same authors also investigated the feasibility of simultaneous material decomposition of three contrast agents (bismuth, iodine, and gadolinium) in vivo in a canine model [59]. They observed tissue enhancement at multiple phases in a single CT acquisition, opening the potential to replace multiphase CT scans by a single CT acquisition with multiple contrast agents (see Fig. 8).

In clinical practice, the use of multi-material maps may be hampered by the unavoidable increase of image noise in a multi-material decomposition. Similar to ultrahigh resolution scanning, nonlinear data and image denoising techniques will

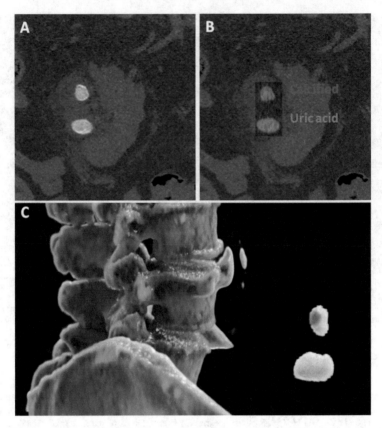

Fig. 7 Combination of high-resolution morphological and functional imaging with photon-counting CT. Abdominal CT scan of a volunteer acquired with the pre-clinical hybrid dual-source CT prototype in "sharp" mode. The high-resolution image at 0.25 mm nominal slice thickness shows the internal structure of two kidney stones (**A**), while a material map reveals their composition (**B**). A volume-rendered image puts both stones into their anatomical context (**C**, note the different color scheme for uric acid and calcium). (Courtesy of S. Leng, Mayo Clinic Rochester, MN, USA)

play a key role in fully exploiting the potential of multi-material decomposition (see, e.g., [60]).

In this review article, we have outlined the basic principles of medical photon-counting CT and its potential clinical applications. Once remaining challenges of this technology have been mastered, photon-counting CT has the potential to bring clinical CT to a new level of performance.

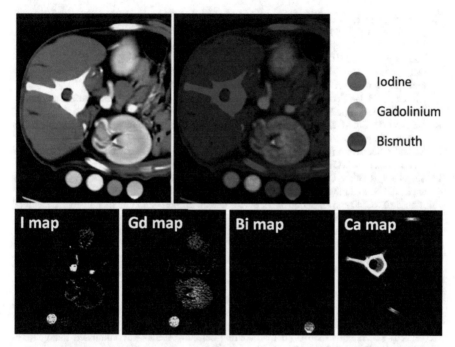

Fig. 8 Simultaneous imaging of three different contrast agents (iodine, gadolinium, and bismuth) by multi-material decomposition in a dog model. Scan data were acquired with the pre-clinical hybrid dual-source CT prototype. Bismuth was administered more than 1 day prior to scanning, followed by intravenous administration of gadolinium-based contrast agent 3–5 min before the scan and iodine-based contrast agent immediately before the scan to visualize different phases of renal enhancement in a single CT acquisition. Scan data were read out in four energy bins (25–50, 50–75, 75–90, and 90–140 keV). Top left: standard CT image combining the data of the four energy bins. The three contrast agents cannot be differentiated. Top right: grayscale image with overlay of the colored material maps. Bottom: the individual material maps. Note the noise amplification by multi-material decomposition. (Courtesy of R Symons, NIH, Bethesda, USA)

References

1. Kalender W, Seissler W, Klotz E, Vock P. Spiral volumetric CT with single-breath-hold technique, continuous transport and continuous scanner rotation. Radiology. 1990;176:181–3.
2. McCollough CH, Zink FE. Performance evaluation of a multi-slice CT system. Med Phys. 1999;26:2223–30.
3. Klingenbeck-Regn K, Schaller S, Flohr T, Ohnesorge B, Kopp AF, Baum U. Subsecond multi-slice computed tomography: basics and applications. Eur J Radiol. 1999;31:110–24.
4. Mori S, Obata T, Nakajima N, Ichihara N, Endo M. Volumetric perfusion CT using prototype 256-detector row CT scanner: preliminary study with healthy porcine model. AJNR Am J Neuroradiol. 2005;26(10):2536–41.
5. Flohr TG, McCollough CH, Bruder H, Petersilka M, Gruber K, Süß C, et al. First performance evaluation of a dual-source CT (DSCT) system. Eur Radiol. 2006;16:256–68.

6. Lu GM, Zhao Y, Zhang LJ, Schoepf UJ. Dual-energy CT of the lung. AJR Am J Roentgenol. 2012;199(5 Suppl):S40–53.
7. Marin D, Boll DT, Mileto A, Nelson RC. State of the art: dual-energy CT of the abdomen. Radiology. 2014;271(2):327–42.
8. Odisio EG, Truong MT, Duran C, de Groot PM, Godoy MC. Role of dual-energy computed tomography in thoracic oncology. Radiol Clin N Am. 2018;56(4):535–48.
9. Albrecht MH, De Cecco CN, Schoepf UJ, et al. Dual-energy CT of the heart current and future status. Eur J Radiol. 2018;105:110–8.
10. De Santis D, Eid M, De Cecco CN, et al. Dual-energy computed tomography in cardiothoracic vascular imaging. Radiol Clin N Am. 2018;56(4):521–34.
11. Rajiah P, Sundaram M, Subhas N. Dual-energy CT in musculoskeletal imaging: what is the role beyond gout? AJR Am J Roentgenol. 2019;213(3):493–505.
12. Siegel MJ, Ramirez-Giraldo JC. Dual-energy CT in children: imaging algorithms and clinical applications. Radiology. 2019;291(2):286–97.
13. Johnson TRC, Krauß B, Sedlmair M, et al. Material differentiation by dual-energy CT: initial experience. Eur Radiol. 2007;17(6):1510–7.
14. Zhang D, Li X, Liu B. Objective characterization of GE discovery CT750 HD scanner: gemstone spectral imaging mode. Med Phys. 2011;38(3):1178–88.
15. Rassouli N, Etesami M, Dhanantwari A, Rajiah P. Detector-based spectral CT with a novel dual-layer technology: principles and applications. Insights Imaging. 2017;8(6):589–98.
16. Feuerlein S, Roessl E, Proksa R, et al. Multienergy photon-counting K-edge imaging: potential for improved luminal depiction in vascular imaging. Radiology. 2008;249(3):1010–6.
17. Taguchi K, Iwanczyk JS. Vision 20/20: single photon-counting x-ray detectors in medical imaging. Med Phys. 2013;40(10):100901.
18. Taguchi K. Energy-sensitive photon-counting detector-based X-ray computed tomography. Radiol Phys Technol. 2017;10(1):8–22.
19. Willemink MJ, Persson M, Pourmorteza A, Pelc NJ, Fleischmann D. Photon-counting CT: technical principles and clinical prospects. Radiology. 2018;289(2):293–312.
20. Leng S, Bruesewitz M, Tao S, et al. Photon-counting detector CT: system design and clinical applications of an emerging technology. Radiographics. 2019;39(3):729–43.
21. Kappler S, Niederlöhner D, Stierstorfer K, Flohr T. Contrast-enhancement, image noise and dual-energy simulations for quantum-counting clinical CT. Proc SPIE Med Imaging Conf. 2010;7622:76223H.
22. Kopp FA, Daerr H, Si-Mohamed S, et al. Evaluation of a pre-clinical photon-counting CT prototype for pulmonary imaging. Sci Rep. 2018;8(1):17386.
23. Bratke G, Hickethier T, Bar-Ness D, et al. Spectral photon-counting computed tomography for coronary stent imaging: evaluation of the potential clinical impact for the delineation of in-stent restenosis. Investig Radiol. 2020;55:61. https://doi.org/10.1097/RLI.0000000000000610.
24. Riederer I, Si-Mohamed S, Ehn S, et al. Differentiation between blood and iodine in a bovine brain – initial experience with spectral photon-counting computed tomography (SPCCT). PLoS One. 2019;14(2):e0212679.
25. Riederer I, Bar-Ness D, Kimm MA. Liquid embolics agents in spectral x-ray photon-counting computed tomography using tantalum K-edge imaging. Sci Rep. 2019;9:5268.
26. Cormode DP, Si-Mohamed S, Bar-Ness D, et al. Multicolor spectral photon-counting computed tomography: in vivo dual contrast imaging with a high count rate scanner. Sci Rep. 2017;7(1):4784.
27. Dangelmaier J, Bar-Ness D, Daerr H, et al. Experimental feasibility of spectral photon-counting computed tomography with two contrast agents for the detection of endoleaks following endovascular aortic repair. Eur Radiol. 2018;28(8):3318–25.
28. Muenzel D, Bar-Ness D, Roessl E, et al. Spectral photon-counting CT: initial experience with dual-contrast agent K-edge colonography. Radiology. 2017;283(3):723–8.
29. Si-Mohamed S, Bar-Ness D, Sigovan M, et al. Multicolour imaging with spectral photon-counting CT: a phantom study. Eur Rad Exp. 2018;2:34.

30. Si-Mohamed S, Tatard-Leitman V, Laugerette A, et al. Spectral Photon-Counting Computed Tomography (SPCCT): in-vivo single-acquisition multi-phase liver imaging with a dual contrast agent protocol. Sci Rep. 2019;9(1):8458.
31. Kappler S, Hannemann T, Kraft E, et al. First results from a hybrid prototype CT scanner for exploring benefits of quantum-counting in clinical CT. In: Medical imaging 2012: physics of medical imaging. San Diego: International Society for Optics and Photonics; 2012. p. 83130X.
32. Kappler S, Henning A, Krauss B, et al. Multi-energy performance of a research prototype CT scanner with small-pixel counting detector. In: Medical imaging 2013: Physics of medical imaging. Lake Buena Vista: International Society for Optics and Photonics; 2013. p. 866800.
33. Kappler S, Henning A, Kreisler B, et al. Photon-counting CT at elevated x-ray tube currents: contrast stability, image noise and multi-energy performance. In: Medical imaging 2014: physics of medical imaging. San Diego: International Society for Optics and Photonics; 2014. p. 90331C.
34. Yu Z, Leng S, Jorgensen SM, et al. Evaluation of conventional imaging performance in a research CT system with a photon-counting detector array. Phys Med Biol. 2016;61:1572–95.
35. Gutjahr R, Halaweish AF, Yu Z, et al. Human imaging with photon-counting-based computed tomography at clinical dose levels: contrast-to-noise ratio and cadaver studies. Investig Radiol. 2016;51(7):421–9.
36. Pourmorteza A, Symons R, Sandfort V, et al. Abdominal imaging with contrast-enhanced photon-counting CT: first human experience. Radiology. 2016;279(1):239–45.
37. Pourmorteza A, Symons R, Reich DS, Bagheri M, Cork TE, Kappler S, Ulzheimer S, Bluemke DA. Photon-counting CT of the brain: in vivo human results and image-quality assessment. AJNR Am J Neuroradiol. 2017;38(12):2257–63.
38. Yu Z, Leng S, Kappler S, et al. Noise performance of low-dose CT_ comparison between an energy integrating detector and a photon-counting detector using a whole-body research photon-counting CT scanner. J Med Imaging. 2016;3(4):043503.
39. Symons R, Cork T, Sahbaee P, et al. Low-dose lung cancer screening with photon-counting CT: a feasibility study. Phys Med Biol. 2017;62(1):202–13.
40. Symons R, Pourmorteza A, Sandfort V, et al. Feasibility of dose-reduced chest CT with photon-counting detectors: initial results in humans. Radiology. 2017;285(3):980–9.
41. Symons R, Sandfort V, Mallek M, Ulzheimer S, Pourmorteza A. Coronary artery calcium scoring with photon-counting CT: first in vivo human experience. Int J Cardiovasc Imaging. 2019;35(4):733–9.
42. Leng S, Rajendran K, Gong H, et al. 150-μm spatial resolution using photon-counting detector computed tomography technology: technical performance and first patient images. Investig Radiol. 2018;53(11):655–62.
43. Symons R, de Bruecker Y, Roosen J, et al. Quarter-millimeter spectral coronary stent imaging with photon-counting CT: initial experience. J Cardiovasc Comput Tomogr. 2018;12:509–15.
44. von Spiczak J, Mannil M, Peters B, et al. Photon-counting computed tomography with dedicated sharp convolution kernels – tapping the potential of a new technology for stent imaging. Investig Radiol. 2018;53(8):486–94.
45. Pourmorteza A, Symons R, Henning A, Ulzheimer S, Bluemke DA. Dose efficiency of quarter-millimeter photon-counting computed tomography: first-in-human results. Investig Radiol. 2018;53(6):365–72.
46. Zhou W, Lane JI, Carlson ML, et al. Comparison of a photon-counting-detector CT with an energy-integrating-detector CT for temporal bone imaging: a cadaveric study. AJNR Am J Neuroradiol. 2018;39(9):1733–8.
47. Bartlett DJ, Koo WC, Bartholmai BJ, et al. High-resolution chest computed tomography imaging of the lungs: impact of 1024 matrix reconstruction and photon-counting detector computed tomography. Investig Radiol. 2019;54(3):129–37.
48. Li Z, Leng S, Yu L, et al. An effective noise reduction method for multi-energy CT images that exploit spatio-spectral features. Med Phys. 2017;44(5):1610–23.
49. Harrison AP, Xu Z, Pourmorteza A, Bluemke DA, Mollura DJ. A multichannel block-matching denoising algorithm for spectral photon-counting CT images. Med Phys. 2017;44(6):2447–52.

50. Rajendran K, Tao S, Abdurakhimova D, Leng S, McCollough C. Ultra-high resolution photon-counting detector CT reconstruction using spectral prior image constrained compressed-sensing (UHR-SPICCS). Proc SPIE Int Soc Opt Eng. 2018;10573:1057318. https://doi.org/10.1117/12.2294628.
51. Leng S, Zhou W, Yu Z, et al. Spectral performance of a whole-body research photon-counting detector CT: quantitative accuracy in derived image sets. Phys Med Biol. 2017;62(17):7216–32.
52. Symons R, Reich DS, Bagheri M, et al. Photon-counting computed tomography for vascular imaging of the head and neck: first in vivo human results. Investig Radiol. 2018;53(3):135–42.
53. Grant KL, Flohr TG, Krauss B, et al. Assessment of an advanced image-based technique to calculate virtual monoenergetic computed tomographic images from a dual-energy examination to improve contrast-to-noise ratio in examinations using iodinated contrast media. Investig Radiol. 2014;49(9):586–92.
54. Zhou W, Abdurakhimova D, Bruesewitz M, et al. Impact of photon-counting detector technology on kV selection and diagnostic workflow in CT. Proc SPIE Int Soc Opt Eng. 2018;10573:105731C. https://doi.org/10.1117/12.2294952.
55. Gutjahr R, Polster C, Henning A, Kappler S, Leng S, McCollough CH, Sedlmair MU, Schmidt B, Krauss B, Flohr TG. Dual-energy CT kidney stone differentiation in photon-counting computed tomography. Proc SPIE Int Soc Opt Eng. 2017;10132:844–50.
56. Ferrero A, Gutjahr R, Halaweish AF, Leng S, McCollough CH. Characterization of urinary stone composition by use of whole-body, photon-counting detector CT. Acad Radiol. 2018;25(10):1270–6.
57. Marcus RP, Fletcher JG, Ferrero A, et al. Detection and characterization of renal stones by using photon-counting-based CT. Radiology. 2018;289(2):436–42.
58. Symons R, Cork TE, Lakshmanan MN, et al. Dual-contrast agent photon-counting computed tomography of the heart: initial experience. Int J Cardiovasc Imaging. 2017;33:1253–61.
59. Symons R, Krauss B, Sahbaee P, et al. Photon-counting CT for simultaneous imaging of multiple contrast agents in the abdomen: an in vivo study. Med Phys. 2017;44(10):5120–7.
60. Tao S, Rajendran K, McCollough CH, Leng S. Material decomposition with prior knowledge aware iterative denoising (MD-PKAID). Phys Med Biol. 2018;63(19):195003.

Coronary Artery Calcifications Assessment with Photon-counting Detector Computed Tomography

Pierre-Antoine Rodesch, Niels R. van der Werf, Salim A. Si-Mohamed, and Philippe C. Douek

1 Introduction

According to the World Health Organization (WHO), cardiovascular diseases (CVDs) were the leading cause of death with 32% of global deaths in 2019 [1]. CVDs impact the heart and blood vessels and can lead to severe acute events such as heart attacks and strokes. In order to monitor patients at risk, a coronary artery calcium (CAC) scoring evaluation can be performed to assess cardiovascular risk [2, 3]. According to this score, the patient may be included in different risk categories: low, medium, or high. The Agatston score (AS) is a computed tomography (CT) protocol to detect calcified plaques [4]. The outcome is a score reflecting the amount of CAC: none, less than average, average, or greater than average.

When the AS was introduced in 1990, it was computed from an electron-beam CT scanner acquisition. In the beginning of the twenty-first century, this technology was superseded by multi-slice CT enabling either helical or step-and-shoot trajectories. Another major CT breakthrough arose recently with the advent of X-ray photon-counting detectors (PCDs), in pre-clinical and clinical CT

P.-A. Rodesch (✉)
Physics and Astronomy, University of Victoria, Victoria, BC, Canada
e-mail: prodesch@uvic.ca

N. R. van der Werf
Department of Radiology, University Medical Center Utrecht, Utrecht, CX, The Netherlands
e-mail: n.r.vanderwerf@umcutrecht.nl

S. A. Si-Mohamed · P. C. Douek
University Lyon, INSA-Lyon, University Claude Bernard Lyon 1, UJM-Saint Etienne, CNRS, Inserm, Villeurbanne, France

Department of Radiology, Louis Pradel Hospital, Hospices Civils de Lyon, Bron, France
e-mail: salim.si-mohamed@chu-lyon.fr; philippe.douek@chu-lyon.fr

© The Author(s), under exclusive license to Springer Nature Switzerland AG 2023
S. Hsieh, K. (Kris) Iniewski (eds.), *Photon Counting Computed Tomography*,
https://doi.org/10.1007/978-3-031-26062-9_2

scanners [5, 6]. Compared to the indirect conversion occurring in energy-integrating detectors (EIDs), PCDs operate by the direct conversion of photons into electron clouds to produce a pulse height signal. This technology shift improves three aspects of image quality [7, 8]:

- **Smaller pixels**: whereas septa are required between adjacent EID pixels to reduce pixel crosstalk, a PCD module presents a continuous detection area. The interseptal gap in EIDs limits the minimum pixel size in order to maintain acceptable dose efficiency. PCD-CT requires no septa and therefore can be equipped with smaller pixels than EID-CT, leading to higher resolution for the same dose efficiency.
- **Contrast**: indirect conversion produces an analog signal where the intensity is proportional to the integrated incoming photon energy. This gives a relatively larger weight to higher energy photons. By contrast, PCDs output one pulse height peak per photon, leading to a more constant weight for each detected photon. As the X-ray attenuation difference between two materials decreases with increasing energy, the PCD contrast will be higher than EID contrast for the same detected spectrum. This increased contrast is more significant for materials such as calcium, which present greater signals for lower photon energies.
- **Noise**: the pulse height voltage peaks generated from incident photons on the PCD are compared to one or several thresholds and sorted according to their amplitude. By setting the lowest threshold just above the electronic noise level, PCDs can effectively eliminate electronic noise and decrease the noise in the resulting images. This effect is even more significant in low-dose acquisitions where the proportion of electronic noise is relatively higher.

The improvement of these three image quality aspects will impact all image reconstructions. Image reconstruction uses the combined signals from all energy bins to compute the conventional CT image in Hounsfield unit (HU). Furthermore, in the last decade, clinical CT devices have begun to offer the possibility of dual energy (DE) which consists of the acquisition of two spectra with a low and high mean energy. Whereas most DE-CTs require a major change in the CT architecture to operate in spectral mode, a PCD-CT can operate in both conventional and spectral modes by simply activating the thresholds inside the detector to perform multi-energy (ME) measurements. The spectral information from either a DE-CT or a PCD-CT scanner is then used by material decomposition algorithms to perform material identification according to the effective atomic number and the density. This information is used in many clinical applications.

In this chapter, we will first present the AS and its limitations. We will then expose the first results obtained with 500 mm field-of-view (FOV) pre-clinical and clinical PCD-CTs, as well as the comparison with EID-CT. These studies were mainly performed with CT scanners from two manufacturers (References [7, 8]). Finally, we will show preliminary results using spectral information for CAC scoring.

2 The Agatston Score (AS) for CAC Scoring

2.1 Definition of the Agatston Score

The AS protocol is still used as it was originally defined in 1990, which was constrained by the technical capacities of an electron-beam CT (Imatron C100 CT scanner—Imatron Inc.) [9]. It uses the conventional image of a multi-slice CT measured with a step-and-shoot acquisition [10] at a tube potential of 120 kVp. The scan covers the entire heart and is subsequently reconstructed with a 3 mm slice thickness. Groups of voxels exceeding the 130 HU threshold are considered as belonging to a plaque in the coronary arteries. As the HU number depends on the tube voltage, the latter is always kept at 120 kVp in all studies to maintain the threshold coherence.

To compute the AS, groups of voxels of at least 1 square millimeter in size are considered as a plaque area. Each plaque area is then multiplied by a factor according to its maximal HU number (cf. Table 1):

For example, a plaque with an 8 mm^2 area and a maximal HU number of 400 will have a score of $8 \times 4 = 32$. The total AS is the sum of each plaque's weighted score. When no feature is detected above 130 HU, the AS is 0.

The AS was introduced in 1990, and its relevance to assess CVD risk has been demonstrated with numerous studies involving a high number of patients [2, 11]. It has been utilized in long-term research studies which have been running for 20 years [12] showing the correlation between the CVD risk categories and stroke events. This makes it an undeniable tool for CAC scoring with massive amounts of collected data. However, as mentioned above, AS is scored on a protocol with a poor through-plane resolution, as a result of the 3 mm slice thickness reconstructions.

These large slice thickness reconstructions, as well as the high contrast of coronary artery calcium, cause partial volume and/or blooming artifacts. High-density plaques appear to be larger than their physical size in the resulting images. The consequences of these artifacts are illustrated in the following subsections on phantoms used for simulating the calcium scoring clinical task.

2.2 Phantom Inserts for AS Performance Assessment

To assess the reproducibility of the AS in-between CT manufacturers or X-ray detection technologies, several phantom inserts have been developed (Fig. 1). They

Table 1 Multiplication factors used for AS

HU maximal value range	Factor
130 < HU < 199	1
200 < HU < 299	2
300 < HU < 299	3
400 < HU	4

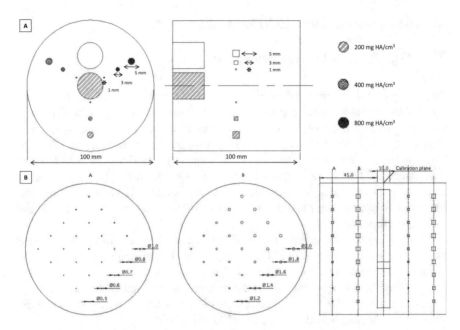

Fig. 1 Diagram of the frontal and side views of the cardiac inserts. (**a**) The CCI insert QRM and (**b**) the D100 insert QRM (QRM, Möhrendorf, Germany, www.qrm.de)

provide a common reference to compare CT modalities [13] by reproducing the human-like contrast of the calcium scoring clinical task using hydroxyapatite (HA) to mimic calcifications in a soft tissue-like plastic background (35 HU) or a water-like background (0 HU). The most common phantom insert is the cardiac calcification insert (CCI, QRM GmbH, Fig. 1a) composed of nine cylindrical calcifications with low (200 mg HA/cm^3), medium (400 mg HA/cm^3), and high (800 mg HA/cm^3) densities with different diameters/length (1, 3, or 5 mm). Another phantom insert is the D100 insert (QRM GmbH, Fig. 1b) containing 100 calcifications of various sizes (from 0.5 to 2.0 mm) and various densities (from 90 to 540 mg HA/cm^3) [14]. These 100 mm diameter inserts are placed in an anthropomorphic thorax phantom to reproduce the global attenuation through the human chest with the possibility to add fat or soft tissue-like surrounding rings to account for patient size variations.

2.3 Impact of the Blooming Artifact

The blooming artifact is a blurring effect due to insufficient resolution. In the case of the AS, the slice thickness is relatively high and creates a blooming artifact in the longitudinal (or axial) direction. This will have two different consequences depending on whether the calcification is small and low density or large and high

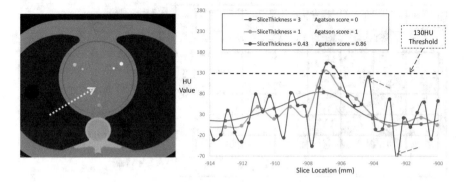

Fig. 2 Left: CCI insert reconstructed with a spectral photon-counting CT (SPCCT; Philips Healthcare) [6], and the yellow arrow points to the profile location. Right: longitudinal profiles through the calcification with low density (200 mg HA/cm^3) and small size (1 mm diameter/length). The AS is indicated in the legend for each case. The theoretical AS would be 2 (1 mm × 290 HU). The dashed arrows point to the noisy data points

density. In order to demonstrate both impacts, we presented the longitudinal profiles plotted in the CCI insert (Fig. 2). In the two cases, we displayed longitudinal profiles for three different conventional images with the same exact reconstruction parameters except for the slice thickness, which was set to 0.43 mm (i.e., isotropic voxel), 1 mm, or 3 mm (i.e., AS protocol).

In Fig. 2, the profiles are plotted through the low-density (200 mg HA/cm^3) small (1 mm diameter/length) calcification of the CCI insert (as shown by the yellow arrow). We observe that the 3 mm thickness case does not produce a voxel with an HU value above the 130 HU threshold, leading to a zero AS. When the resolution is increased, the calcification becomes detectable. However, the HU intensity is still underestimated (a 290 HU value is expected for a 200 mg HA/cm^3 density). And at the higher resolution, we can see that the noisy data points making up the profile can have intensities which exceed the 130 HU threshold and therefore interfere with the AS estimation.

The second effect of the blooming artifact impacts the large and high attenuating plaques. In Fig. 3, the same profiles are displayed as those shown in Fig. 2, but through the high-density (800 mg HA/cm^3, expected CT number: 1190 HU) large calcification (5 mm diameter/length). We can see that the calcification is detected in all cases. But if we look at the associated AS (mentioned above the profiles), there is a significant difference between the 3 mm reconstruction and the two other instances. Indeed, the low reconstructed resolution results in the volume of the plaque to be overestimated, as it appears to have a higher area in the longitudinal direction above the 130 HU threshold. And the maximum attenuation value is similar in the three cases. Moreover, in the 3 mm case, we observe that three slices are above 130 HU. But a small translation in the longitudinal direction could easily lower this number to 2 slices and significantly reduce the AS (33%). This illustrates

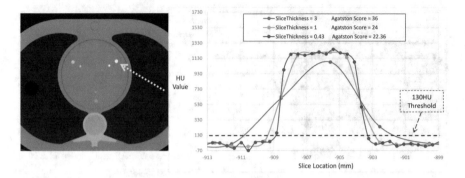

Fig. 3 Left: the CCI insert reconstructed with a spectral photon-counting CT (SPCCT; Philips Healthcare) [6]. The yellow arrow points to the profile location. Right: the longitudinal profiles through the calcification with high density (800 mg HA/cm^3) and large size (5 mm diameter/length). The AS is indicated in the legend for each case. The theoretical AS would be 20 (5 mm × 1190 HU)

that the reproducibility of the AS decreases with the resolution of the image and the AS can vary according to patient position [15].

We have presented in this subsection the double effect of decreasing resolution: the smallest and lowest density plaques are not detected, and the volume of largest and most dense plaques is overestimated. Both of these effects will be further explored in the following section comparing EID-CT and PCD-CT.

3 Comparison of EID-CT and PCD-CT for Calcium Scoring

In this section, we will compare the EID-CT and PCT-CT results that have been published in phantom studies. We refer to each study according to the following rule: a contraction of the first author's name and publication year (e.g., "Symons et al. [16]"). All cited articles that include both modalities compare a prototype PCD-CT with a state-of-the-art EID-CT of the same manufacturer. This provides a fair comparison but is constrained by the limits related to the prototype devices. Two different CT manufacturers are cited in this chapter (Table 2).

The studies with measurements on the phantom described in subsection II.b always included 5 repetitions with small translations and rotations of the inserts between each repetition to consider the variability caused by the object position. Repetitions enable the evaluation of the statistical p value. We define a difference between EID and PCD as significant when $p < 0.05$. We have summarized the results in four categories: AS values, noise reduction, resolution, and dose robustness.

Table 2 Bibliography table of cited reports for the comparison between EID-CT and PCD-CT for the Agatston score evaluation with a conventional image

Manufacturer	Study
Siemens Healthcare, Forchheim, Germany	Symons et al. [16], Sandstedt et al. [17], Eberhard et al. [18]
Philips Healthcare, Haifa, Israel	VanderWerf et al. [19, 20]

3.1 Agatston Score Values

Several works have compared the AS between an EID-CT and a PCT-CT: Symons et al. [16], Eberhard et al. [18], and VanderWerf et al. [20]. At matching doses, no significant differences between the modalities were observed in any of the studies. However, VanderWerf et al. [20] demonstrated that when diminishing the slice thickness or the slice increment, the PCD-CT presented a statistically significant reduction in the AS for large, dense calcifications, confirmed by a decrease of the detected volume.

3.2 Noise Reduction

All articles that reported noise measurements in a homogeneous background material showed noise reduction for PCDs. The background was either water or a water-like plastic (around 35 HU). The values of the noise decrease at matching doses were (as a percentage reduction of the EID value): Symons et al. [16] (3.9%), Sandstedt et al. [17] (7.3%), VanderWerf et al. [20] (5.3%), and VanderWerf et al. [19] (5.4%). Another calcium scoring study with a PCD flat panel mounted on a laboratory bench (Juntunen et al. [21]) presented a 11.4% noise reduction compared to a clinical EID.

These observations are in line with research works comparing EID-CT and PCD-CT, including those involving human patients on different clinical protocols: head and neck [22], lung imaging [6, 23], coronary CT angiography [24, 25], abdominal CT [26], etc.

Although one study reported an increase in contrast for PCDs (Symons2019), no global prevailing trend exists for HU number difference in calcium plaques in conventional images.

3.3 Resolution Increase

An image quality metric to measure resolution is the computation of the modulation transfer function (MTF). Technical reviews of large field-of-view PCD-CTs [5, 6] reported better MTF values for PCD-CTs without specifying the exact clinical protocols. As demonstrated in Sect. 2.3, this will have several consequences on the AS.

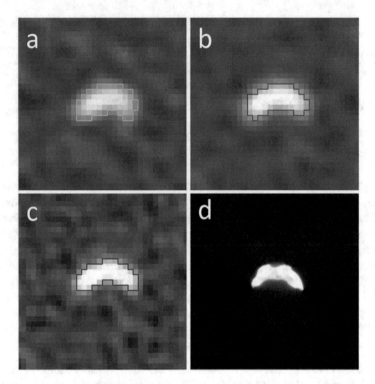

Fig. 4 CT slices of a calcification in a cadaveric heart from Sandstedt et al. [17] (display window/level: 2128/458 HU). The overlay is the voxels above 130 HU: (**a**) with an EID-CT (62 voxels), (**b**) with a PCD-CT using the same kernel as the EID-CT: 55 voxels, (**c**) with a PCD-CT using a sharper kernel than **a** and **b** (48 voxels), and (**d**) the corresponding micro-CT image (the same scale)

First, VanderWerf et al. [20] and VanderWerf et al. [19] have demonstrated that for the same voxel size, more calcifications were detected in the D100 insert with a PCD-CT than with an EID-CT (Fig. 4). The number of calcifications that were not seen by the EID-CT but were detected with the PCD-CT increased when diminishing the slice thickness. This means that in some cases a zero AS with an EID could be positive with a PCD-CT.

Second, some studies have investigated the calcification volume measurement. Compared to the ground truth value, CT images always overestimate the volume, especially for large, dense plaques. VanderWerf et al. [19, 20], and Sandstedt et al. [17] have shown PCT-CT reduced this overestimation for matching dose and reconstruction parameters. Sandstedt et al. [17] has also reported further improved volume estimations with a PCD-CT by using sharper reconstruction parameters not available on an EID-CT (Fig. 5).

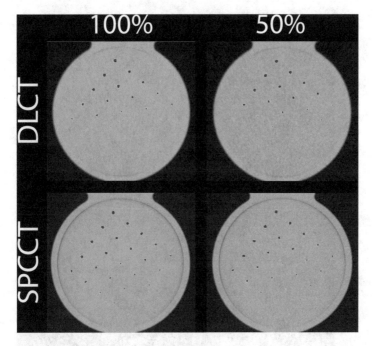

Fig. 5 A slice of the D100 insert measured with either an EID-CT (DLCT: Dual-Layer CT, IQon Spectral CT, Philips Healthcare) or a PCD-CT (Spectral Photon-Counting CT, Philips Healthcare) at 100% (left) or 50% (right) of the recommended dose. The red overlay indicates the 130 HU threshold

3.4 Dose Robustness

The last comparison point is the dose robustness of each modality, i.e., how results degrade when decreasing the irradiation dose. As the AS is defined with a 120 kVp tube voltage, dose reduction was achieved by lowering the tube current. Several authors investigated dose reduction at different levels. Symons et al. [16] explored a 75% dose reduction for EID-CT and PCD-CT (Fig. 6). They demonstrated that the AS deteriorated compared to the normal dose with EID-CT while the PCD-CT results remained in agreement with the normal dose. VanderWerf et al. [19] investigated a 50% dose reduction for both EID-CT and PCD-CT without any significant differences of the AS.

3.5 Discussion and Limits

In conclusion, the improved spatial resolution of the PCD-CT reduced the blooming artifact. This translated into the detection of smaller calcifications and a decrease in

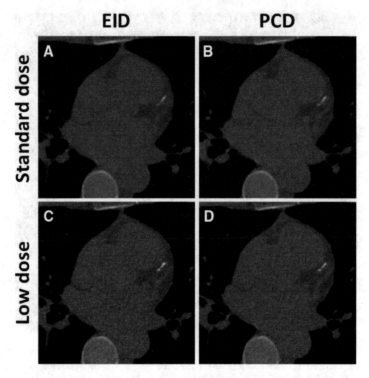

Fig. 6 Example case of 67-year-old man [16]. The low-dose images were obtained by diminishing the tube current by 75% from the standard protocol

the volume overestimation. However, this did not significantly impact the overall AS values due to the strong constraints of this protocol (3 mm slice thickness and 120 kVp tube voltage). In the same way the AS protocol with multi-slice CT was designed to provide coherent results with the vastly different electron-beam CT [27], we expect the results to be coherent between EID-CT and PCD-CT, as the differences between those modalities are not significant in comparison.

However, several studies have experimented to modify the AS protocol to improve reproducibility and detection. They concluded that there are significant differences with PCD-CT enabling them to maintain reproducibility and acceptable noise levels when decreasing the dose or the slice thickness.

The published methods and the conclusions we have presented have several limits. Firstly, only two of them used real patient acquisitions or dynamic phantoms. This means that the impact of patient movement was not considered in most of the cited articles. Secondly, most of the conclusions were drawn without investigating the variation caused by the studied object size. This could influence results because of beam-hardening artifacts. It has also been demonstrated that the change in patient size can cause more variation than the change of scanner model [13]. VanderWerf2021a and VanderWerf2022a have included two different phantom sizes and have shown better performances for PCD-CT when increasing the size. However, there is not sufficient data to demonstrate a global trend. Finally, the cited studies followed the manufacturer recommendations for the iterative reconstruction (IR) level, which has an influence on calcium scoring [28]. Higher IR levels can be used to maintain a low noise level while either reducing irradiation dose for EID-CT or using sharper reconstruction parameters for PCD-CT.

The results in this section were based on conventional CT images. Another key point of PCD-CT is to operate in spectral mode without major CT geometry modifications. The next section presents the preliminary results using the PCD spectral information for calcium scoring.

3.6 Clinical Implications

The AS has important clinical consequences according to its level. In the case of a medium (>100) or high AS (>300), the patient is identified as at risk. According to the intensity of the AS, further clinical procedures are initiated. As reported in several studies, the AS levels are statistically similar between EID-CT and PCD-CT. However, PCD-CT could enable further dose reduction for this protocol while maintaining reproducibility.

In the case of a zero or low AS level (<100) and for patients older than 40 years, the CAC score will play a major role in the decision to start lifelong preventive therapies. Indeed, it has been demonstrated that a zero AS is a strong predicator of a low 10-year CVD risk compared to a low AS [29, 30]. This means that for patients who have been classified as at risk based on global assessments (hypertension, diabetes, CVD familial history), delaying preventive therapy could be considered in the case of a zero AS [31]. An AS evaluation is the most reliable risk marker to downgrade risk classification [32, 33]. However, as demonstrated in Sect. 2.3, a zero AS does not necessarily imply the absence of calcium plaques [34]. Coronary CT angiography (CCTA) performed for zero AS revealed the presence of mixed or calcified plaques in 0.8% [35] or 4% [36] of cases. PCD-CT presents an improved resolution and an ability to detect small and low-density calcifications. It could provide stronger statistical differences between a zero and low AS which would offer significant benefits in diagnosis.

4 Spectral Information for Calcium Scoring

4.1 Virtual Monoenergetic Images (VMIs)

The spectral information provided by a spectral scanner can be used to generate different types of images according to the clinical need [37]. One of them is the virtual monoenergetic images which have been exploited in several published studies performed with a PCD-CT [18, 38–40] to evaluate the AS. VMIs are synthetic images mimicking a CT image acquired with a true monoenergetic photon source between 40 and 200 keV. This corrects for the beam-hardening artifacts present in the conventional image. The contrast and the noise in the image depend on the selected energy.

We display in Fig. 7 the VMI at 50 keV, at 100 keV, and the conventional image of the CCI insert. We can see on the profiles that the contrast decreases with the energy and then the contrast in the conventional image corresponds to a VMI between 60 keV and 80 keV, depending on the IR level selected. Moreover, we can see the VMI at 50 keV presents a higher noise level. The noise increase in lower energies is more important than the contrast enhancement at low energies. This means the optimal energy for CNR is between 60 and 80 keV, as corroborated by several studies [18, 37, 41]. On EID dual-energy CTs, VMIs have been demonstrated to provide a better CNR compared to conventional imaging, and VMI reconstruction has been recommended for clinical routine [42].

The use of VMIs computed from a PCD-CT for AS evaluation has been investigated by several studies, all using the same CT manufacturer. They have all concluded that VMIs around 70 keV are suitable for AS measurement. Skoog et al. [38] have demonstrated the VMI at 72 keV with a low IR level provided the best AS correlation with an EID-CT. They have studied cadaveric hearts and demonstrated the spatial reproducibility of their measurement with two different

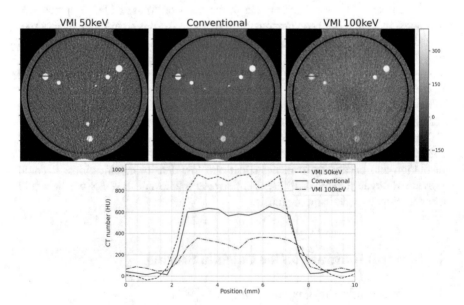

Fig. 7 Top row: VMI at 50 keV, conventional and VMI at 100 keV images of the CCI inserts reconstructed with a spectral photon-counting CT (SPCCT; Philips Healthcare) [6]. VMIs were not computed with a noise-optimized algorithm as available on most clinical CTs. Bottom row: profile through the large medium density calcifications represented in the images with a red line

positions. Eberhard et al. [18] have found the optimal energy for the AS correlation with an EID-CT to be 65 or 70 keV depending on the selected IR level. They have shown the AS decreases when the IR level is raised, impacting the correlation with the reference AS. Van der Werf et al. [39] have demonstrated VMIs at 74 and 76 keV to have comparable AS with an EID-CT and a moving phantom. Van der Werf et al. [40] have computed VMIs at 70 keV from acquisitions with different tube voltages (90, 100, and 120 kVp). They have demonstrated the reproducibility of the AS computed from 90 or 100 kVp measurements compared to the 120 kVp reference. This could enable to reduce the radiation dose, by not only reducing the exposure but also the tube voltage, which cannot be achieved with a conventional image.

In conclusion, VMIs have shown promising results for AS evaluation. A VMI computed at about 70 keV provides the same AS levels as a 120 kVp conventional image. Additionally, improved CNR for the calcification detection task has been reported at energies lower than 70 keV on DE-CT [22] and PCD-CT [40]. It could be used to further improve low-density calcifications detection. However, VMIs computed with PCD-CT present an image quality inferior to conventional imaging in recent studies [41, 43]. Dedicated noise and artifact reduction algorithms should be soon available. Finally, VMIs are independent of the tube voltage used for the acquisition and could improve inter-patient size variability [42]. Another way to exploit the spectral information is to perform material classification as we show in the next subsection.

4.2 Material Classification

Material classification can be performed from at least two images at different energies. For example, the linear attenuation ratio (LAR) between two energies is a function of the effective atomic number and can be used to discriminate between different materials [37]. We display in Fig. 8 the LAR computed from the VMI at 50 and 100 keV for the three largest calcifications of the CCI insert. We represent the measurement with an ellipse whose center is the mean, and the semi-axis is the standard deviation along the x- and y-axis. The values for the background water-like material are represented as well. We can see the calcifications at different HA densities present a constant LAR with an incertitude caused by the noise. The corresponding area does not intersect with the background material which has a different LAR. This area figures the HA material identification.

This spectral "footprint" of the calcifications could be clinically relevant to discriminate low-density calcification from soft tissue material with a close CT number in the conventional images. It could also be used to evaluate the density of calcifications and to improve patient risk classification by proposing a calcification distribution over size and density [44].

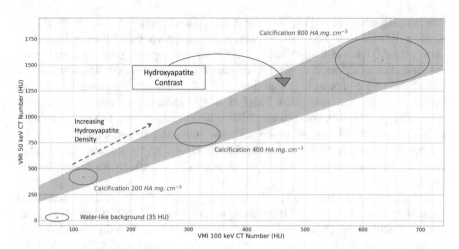

Fig. 8 CT number measurements of the three large calcifications at low (200 mg HA/cm³), medium (400 mg HA/cm³), and large (800 mg HA/cm³) densities. The CT number is represented in a dual-energy space where the x-axis and y-axis are, respectively, the CT numbers of the VMI at 100 keV and 50 keV. The numbers have been measured in the images presented in Fig. 7

5   Conclusion

The Agatston score (AS) is a powerful diagnostic tool for CVD classification. PCD-CT has been shown to provide the same AS levels as EID-CT for the standard AS protocol. However, when upgrading the AS protocol by either reducing the dose or the slice thickness, PCD-CT presents better performance than EID-CT regarding the reproducibility or the background noise requirements. PCD-CT presents higher spatial resolution than EID-CT and can operate in ultra-high-resolution modes not available with standard EID-CT. Additionally, the spectral information measured by PCD-CT can be exploited to generate VMIs around 70 keV, which provide comparable AS levels as a conventional image. CAC scoring is also subject to evolution with investigation of new reduced dose alternatives of the standardized approach. Several protocols have been proposed with a 1 mm slice thickness and 100 kVp acquisition [44] or different scoring methods (volume or mass score) [13, 44]. Spectral information could also be further exploited for calcification characterization.

References

1. World Health Organization (2021) Cardiovascular diseases (CVDs). https://www.who.int/news-room/fact-sheets/detail/cardiovascular-diseases-(cvds). Accessed 15 Mar 2022
2. Detrano R, Guerci A, Carr J, et al. Coronary calcium as a predictor of coronary events in four racial or ethnic groups. N Engl J Med. 2008;358:1336–45

3. Budoff M, Achenbach S, Blumenthal R, et al. Assessment of coronary artery disease by cardiac computed tomography: a scientific statement from the American Heart Association Committee on Cardiovascular Imaging and Intervention, Council on Cardiovascular Radiology and Intervention, and Committee on Cardiac Imaging, Council on Clinical Cardiology. Circulation 2006;114:1761–91

4. Alluri K, Joshi PH, Henry TS, et al. Scoring of coronary artery calcium scans: history, assumptions, current limitations, and future directions. Atherosclerosis 2015;239(1):109–17

5. Rajendran K, Petersilka M, Henning A, et al. First clinical photon-counting detector CT system: technical evaluation. Radiology. 2021. https://doi.org/10.1148/radiol.212579

6. Si-Mohamed S, Boccalini S, Rodesch PA, et al. Feasibility of lung imaging with a large field-of-view spectral photon-counting CT system. Diagn Interv Imaging. 2021;102(5):305–12

7. Danielsson M, Persson M, Sjölin M. Photon-counting X-ray detectors for CT. Phys Med Biol. 2021;66(3):03TR01

8. Si-Mohamed S, Miailhes J, Rodesch PA, et al. Spectral photon-counting CT technology in chest imaging. J Clin Med. 2021;10,5757

9. Agatston A, Janowitz W, Hildner H, et al. Quantification of coronary artery calcium using ultra-fast computed tomography. J Am Coll Cardiol. 1990;15(4):827–32

10. Hsieh J, Londt J, Vass M, et al. Step-and-shoot data acquisition and reconstruction for cardiac X-ray computed tomography. Med Phys. 2006;33(11):4236–48

11. Greenland P, Blaha M, Budoff M, et al. Coronary calcium score and cardiovascular risk. J Am Coll Cardiol. 2018;72(4):434–47

12. Criqui M, Denenberg J, Ix J, et al. Calcium density of coronary artery plaque and risk of incident cardiovascular events. JAMA 15 2014;311(3): 271–78

13. McCollough C, Ulzheimer S, Halliburton S. Coronary artery calcium: a multi-institutional, multimanufacturer international standard for quantification at cardiac CT. Radiology 2007;243(2):527–38

14. Groen JM, Kofoed KF, Zacho M, et al. Calcium score of small coronary calcifications on multidetector computed tomography: results from a static phantom study. Eur J Radiol. 2008;82(2):e58–63

15. Rutten A, Isgum I, Prokop M. Coronary calcification: effect of small variation of scan starting position on Agatston, volume, and mass scores. Radiology 2019;246(1):90–8

16. Symons R, Sandfort V, Mallek M, et al. Coronary artery calcium scoring with photon-counting CT: first in vivo human experience. Int J Cardiovasc Imaging. 2019;35(4):733–39

17. Sandstedt M, Marsh J, Rajendran K, et al. Improved coronary calcification quantification using photon-counting-detector CT: an ex vivo study in cadaveric specimens. Eur Radiol. 2021;31(9):6621–30

18. van der Werf NR, Si-Mohamed S, Rodesch PA, et al. Coronary calcium scoring potential of large field-of-view spectral photon-counting CT: a phantom study. Eur Radiol. 2022;32(1):152–62

19. van der Werf NR, Rodesch PA, Si-Mohamed S, et al. (2022) Improved coronary calcium detection and quantification with low-dose full field-of-view photon-counting CT: a phantom study. Eur Radiol. https://doi.org/10.1007/s00330-021-08421-8

20. Juntunen MAK, Kotiaho AO, Nieminen MT, et al. Optimizing iterative reconstruction for quantification of calcium hydroxyapatite with photon counting flat-detector computed tomography: a cardiac phantom study. J Med Imaging. 2021;8(5):052102

21. Symons R, Reich DS, Bagheri M, et al. Photon-counting computed tomography for vascular imaging of the head and neck: first in vivo human results. Invest Radiol. 2018;53(3):135–42

22. Si-Mohamed SA, Greffier J, Miailhes J, et al. Comparison of image quality between spectral photon-counting CT and dual-layer CT for the evaluation of lung nodules: a phantom study. Eur Radiol. 2022;32(1):524–32

23. Rotzinger DC, Racine D, Becce F, et al. Performance of spectral photon-counting coronary CT angiography and comparison with energy-integrating-detector CT: objective assessment with model observer. Diagnostics. 2021;11:2376

24. Si-Mohamed SA, Boccalini S, Lacombe H. Coronary CT angiography with photon-counting CT: first-in-human results. Radiology. 2022. https://doi.org/10.1148/radiol.211780
25. Higashigaito K, Euler A, Eberhard M, et al. Contrast-enhanced abdominal ct with clinical photon-counting detector CT: assessment of image quality and comparison with energy-integrating detector CT. Acad Radiol. 2021. S1076-6332(21)00305-6
26. Ulzheimer S, Kalender WA. Assessment of calcium scoring performance in cardiac computed tomography. Eur Radiol. 2003;13(3):484–97
27. van der Werf NR, Willemink MJ, Willems TP, et al. Influence of iterative reconstruction on coronary calcium scores at multiple heart rates: a multivendor phantom study on state-of-the-art CT systems. Int J Cardiovasc Imaging. 2020;34(6):947–957
28. McCollough C, Boedeker, K, Cody D, et al. Principles and applications of multienergy CT: report of AAPM task group 291. Med Phys. 2020;47:e881–912
29. Skoog S, Henriksson L, Gustafsson H, et al. Comparison of the Agatston score acquired with photon-counting detector CT and energy-integrating detector CT: ex vivo study of cadaveric hearts. Int J Cardiovasc Imaging. 2022. https://doi.org/10.1007/s10554-021-02494-8
30. Eberhard M, Mergen V, Higashigaito K, et al. Coronary calcium scoring with first generation dual-source photon-counting CT-first evidence from phantom and in-vivo scans. Diagnostics. 2021;18;11(9):1708
31. van der Werf NR, Booij R, Greuter MJ, et al. Reproducibility of coronary artery calcium quantification on dual-source CT and dual-source photon-counting CT: a dynamic phantom study. Int J Cardiovasc Imaging. 2022. https://doi.org/10.1007/s10554-022-02540-z
32. van der Werf NR, van Gent M, Booij R, et al. Dose reduction in coronary artery calcium scoring using mono-energetic images from reduced tube voltage dual-source photon-counting CT data: a dynamic phantom study. Diagnostics. 2021;11:2192
33. Chappard C, Abascal J, Olivier C, et al. Virtual monoenergetic images from photon-counting spectral computed tomography to assess knee osteoarthritis. Eur Radiol Exp. 2022;6:10
34. Lam S, Gupta R, Levental M, et al. Optimal virtual monochromatic images for evaluation of normal tissues and head and neck cancer using dual-energy CT. AJNR 2015;36(8);1518–24
35. Tota-Maharaj R, Al-Mallah MH, Nasir K, et al. Improving the relationship between coronary artery calcium score and coronary plaque burden: addition of regional measures of coronary artery calcium distribution. Atherosclerosis 2015;238(1):126–31
36. Willemink M, van der Werf NR, Nieman K, et al. Coronary artery calcium: a technical argument for a new scoring method. J Cardiovasc. 2018;13(6):347–352
37. van Praagh GD, Wang J, van der Werf NR, Greuter MJW, Mastrodicasa D, Nieman K, van Hamersvelt RW, Oostveen LJ, de Lange F, Slart RHJA, Leiner T, Fleischmann D, Willemink MJ. Coronary artery calcium scoring: toward a new standard. Invest Radiol. 2022;57(1):13–22. https://doi.org/10.1097/RLI.0000000000000808
38. Skoog S, Henriksson L, Gustafsson H, Elvelind S, Persson A. Comparison of the Agatston score acquired with photon-counting detector CT and energy-integrating detector CT: ex vivo study of cadaveric hearts. Int J Cardiovasc Imaging. 2022;38(5):1145–55. https://doi.org/10.1007/s10554-021-02494-8
39. van der Werf NR, Booij R, Greuter MJW, Bos D, van der Lugt A, Budde RPJ, van Straten M. Reproducibility of coronary artery calcium quantification on dual-source CT and dual-source photon-counting CT: a dynamic phantom study. Int J Cardiovasc Imaging. 2022;38(7):1613–9. https://doi.org/10.1007/s10554-022-02540-z
40. van der Werf NR, van Gent M, Booij R, Bos D, van der Lugt A, Budde RPJ, Greuter MJW, van Straten M. Dose reduction in coronary artery calcium scoring using mono-energetic images from reduced tube voltage dual-source photon-counting CT data: a dynamic phantom study. Diagnostics. 2021;11(12). https://doi.org/10.3390/diagnostics11122192
41. Michael, Arwed Elias and Boriesosdick, Jan and Schoenbeck, Denise and Woeltjen, Matthias Michael and Saeed, Saher and Kroeger, Jan Robert and Horstmeier, Sebastian and Lennartz, Simon and Borggrefe, Jan and Niehoff, Julius Henning, Image-quality assessment of Polyenergetic and virtual Monoenergetic reconstructions of unenhanced CT scans of the head: initial experiences with the first photon-counting CT approved for clinical use. Diagnostics 2022;12(2): 265. https://doi.org/10.3390/diagnostics12020265

42. Albrecht MH, Vogl TJ, Martin SS, Nance JW, Duguay TM, Wichmann JL, de Cecco CN, Varga-Szemes A, van Assen M, Tesche C, Joseph Schoepf U. Review of clinical applications for virtual monoenergetic dual-energy CT. Radiology. 2019;293(2):260–71. https://doi.org/10.1148/radiol.2019182297
43. Chappard C, Abascal J, Olivier C, Si-Mohamed S, Boussel L, Baptiste Piala J, Douek P, and Peyrin F, Virtual monoenergetic images from photon-counting spectral computed tomography to assess knee osteoarthritis. European Radiology Experimental. 2022;6(1). https://doi.org/10.1186/s41747-021-00261-x
44. Willemink MJ, van der Werf NR, Nieman K, Greuter MJW, Koweek LM, Fleischmann D. Coronary artery calcium: a technical argument for a new scoring method. J Cardiovasc Comput Tomogr. 2019;13(6):347–52. https://doi.org/10.1016/j.jcct.2018.10.014

MARS for Orthopaedic Pathology

**Jennifer A. Clark, Krishna M. Chapagain, Maya R. Amma,
Mahdieh Moghiseh, Chiara Lowe, and Anthony P. H. Butler**

1 Introduction

In 2020 the MARS Extremity 5 × 120 spectral photon-counting CT (SPCCT) was developed by MARS Bioimaging Ltd (MBI). This dedicated upper extremity imaging system aims to address the limitations of conventional imaging modalities used in orthopaedics. The SPCCT technique embodied in the MARS imaging system uses photon processing and charge summing to enable high-resolution (<0.1 mm) imaging of cortical and trabecular bone, imaging of the bone-metal interface with minimal artefact and soft tissue differentiation and information pertaining to bone quality through the use of material identification and quantification. Furthermore, MARS imaging systems are compact scanners designed for point-of-care, community-based imaging to support hospitals in meeting a greater demand for diagnostic services. Excellent imaging characteristics combined with improved

J. A. Clark · K. M. Chapagain · M. Moghiseh (✉)
MARS Bioimaging Limited, Christchurch, New Zealand

University of Otago Christchurch, Christchurch, New Zealand
e-mail: mahdieh.moghiseh@marsbioimaging.com

M. R. Amma · C. Lowe
MARS Bioimaging Limited, Christchurch, New Zealand

A. P. H. Butler
MARS Bioimaging Limited, Christchurch, New Zealand

University of Otago Christchurch, Christchurch, New Zealand

University of Canterbury, Christchurch, New Zealand

European Organisation for Nuclear Research (CERN), Geneva, Switzerland

Human Interface Technology Laboratory New Zealand, University of Canterbury, Christchurch, New Zealand patient accessibility should result in timely imaging and ultimately contribute to successful long-term outcomes for orthopaedic patients.

2 Background

2.1 Conventional Imaging Modalities for Orthopaedic Applications

Plain radiography (X-ray) is commonly used as a first-line imaging modality for acute orthopaedic injuries. It has high accessibility, low cost, low radiation dose and high spatial resolution. Plain radiography is however limited by the superimposition of bony structures, particularly in the wrist, and fractures can remain radiographically occult [1].

Computed tomography (CT) and cone beam CT (CBCT) are used for imaging complex fractures, to assess bone healing, for pre-operative planning and for post-surgical assessment of metal implants. They provide cross-sectional imaging of cortical and trabecular bone with good spatial resolution (1–2 mm). Multiplanar reconstructions allow the assessment of rotational alignment and provide further detail for assessing callus formation and non-union, two features that are under-appreciated on plain X-ray [2]. CBCT can be available in compact scanners for extremity scanning at some regional and community radiology clinics. Recent studies by [3, 4] show that CBCT is superior to plain X-rays for the diagnosis of acute scaphoid fractures which suggests it could be considered for first presentation imaging if the scanner is available. The limitations of standard CT and CBCT include low sensitivity (compared to bone scintigraphy), artefacts arising from metal implants and limited information pertaining to bone quality. For example, it is not able to quantify bone mineral density in the clinical setting. CT also has limited soft tissue differentiation and can fail to demonstrate bone bruising or bone oedema which could delay diagnosis and treatment [5].

Dual energy CT can provide additional material and soft tissue information compared to conventional CT techniques by imaging using two different energy channels. Some of the challenges with dual energy CT are the inability to demonstrate oedema adjacent to cortical bone and oedema/infiltration in the regions where higher proportions of red marrow are present [6]. Furthermore, the presence of sclerosis, for example, with degenerative pathologies, and air in the imaging volume can lead to false-positive results for oedema detection [7].

Other high-resolution CT techniques such as high-resolution peripheral quantitative computed tomography (HRpQCT) are also limited in the vicinity of implants [8]. Therefore, invasive and destructive techniques such as histology or electron microscopy are employed for the assessment of osseous integration in the research setting.

MRI has been shown to have similar diagnostic value to CT with regard to diagnosing occult fractures [9–11] and can be used to detect bone oedema because it provides excellent soft tissue contrast [12]. However, MRI has long scan times which can affect patient compliance and motion artefact and poor spatial resolution (3–4 mm), and it cannot be used for patients with ferromagnetic metallic implants. Furthermore, the infrastructure and cost required to install latest-generation MRI systems can be a significant barrier to the availability of this technology in regional hospitals and smaller medical centres [13].

Radionuclide bone scanning or bone scintigraphy is also a valuable tool for fracture diagnosis and imaging of bone turnover [9]. Bone scintigraphy is a form of nuclear imaging which uses radioactive pharmaceutical agents to highlight increased metabolic activity. For example, areas of rapid bone growth display as regions of high uptake of the radiopharmaceutical ("hot spots"), while areas of bone loss demonstrate very little or no uptake. Although bone scintigraphy has greater sensitivity compared to CT and MRI for the detection of fractures, the technique involves the injection of a radionuclide, exposes the patient to a higher level of radiation and has a diagnostic delay not less than 72 hours [9]. In addition, specialised technicians are required to perform the scan.

For the imaging of bone health and specifically, bone mineral content, dual energy X-ray absorptiometry (DEXA) is the current gold standard. This system estimates areal bone mineral density as the marker of bone strength by utilising the difference in X-ray attenuation for lower and higher energies of X-ray spectrum [14, 15]. DEXA imaging is a two-dimensional bone health assessment method; therefore, this method has several limitations. Firstly, measurement is inaccurate in the presence of high-attenuating objects such as metal implants. Secondly, the assessment is devoid of structural information of the bone and cannot provide information regarding a particular region of a bone, such as monitoring the outcomes of a therapeutic intervention [16, 17]. In addition, DEXA measures abnormal bone mineral values relative to a healthy reference population, and it cannot provide the proportion of bone mineralisation on an individualised basis. Therefore, people classified as having normal bone mineral values in DEXA may still be at risk of developing fractures.

2.2 Spectral Photon-Counting CT with the MARS Imaging System

The MARS imaging systems are spectral photon-counting CT systems. The MARS Extremity 5 × 120, shown in Fig. 1, is specifically designed for point-of-care upper limb imaging.

A standard X-ray tube with a tungsten anode is used to produce a polychromatic X-ray beam. The X-ray beam is attenuated as it passes through the patient and the transmission data is captured by the X-ray camera. The small focal spot of the X-ray

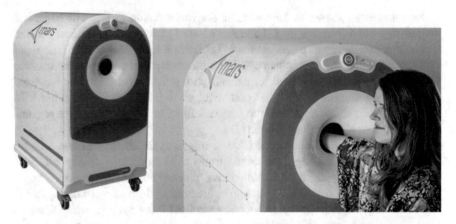

Fig. 1 Left: MARS Extremity 5 × 120 Scanner. Right: Patient seated with limb placed through the entry port

tube and the 0.11 mm square pixels in the camera provide images with high spatial resolution. The MARS imaging system also provides high-energy resolution as it captures attenuation as a function of energy.

It has been known for over 100 years that the X-ray absorption spectrum for atomic elements is unique and depends on the atomic number, Z, and the density of the element [18]. X-ray photons used in medical diagnosis typically have photon energies from 20 to 130 keV [19]. The K-shell and L-shells of atoms contribute most to the photo-electric part of the absorption spectrum [20]. The camera registers the energy of individual X-ray photons in the semiconductor sensor chips and allocates each photon to a protocol-defined energy bin. By keeping track of the number of photons in each energy bin, the camera records different linear attenuation profiles for each X-ray path through the patient. This approach enables attenuation images to be generated for different energy ranges using one X-ray source and one exposure. Subsequent conversion of energy images into material images, with associated material density measures, enables visualisation and analysis of the chemical composition of body materials in the various anatomical and pathological structures. In the diagnostic energy range, it is possible to differentiate between contrast from photo-electric effect, Compton scatter and certain metal k-edges. These can be processed to determine bone mineral content and soft tissue content.

The MARS camera has a two-layer detector. The first layer, the sensor layer, is the semiconductor that converts the X-ray photon to an electron-hole charge cloud in the semiconductor. An electric field of around 500 V/mm drives the charge cloud to electrodes on the top and bottom of the sensor "chip". The electrodes convey the charge from each single X-ray photon as a voltage pulse to the second layer, the application specific integrated circuit (ASIC) chip. The ASIC has multiple electrical pulse-shaping circuits that measure and record the size of the pulses. The ASIC in this device is the Medipix3RX ASIC [21]. Its patented circuitry enables inter-pixel charge sharing effects, measures each pulse height (proportional to the X-ray photon's energy) and allocates a count into the relevant energy bins [22].

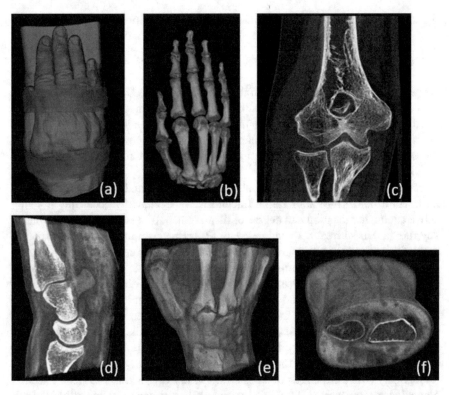

Fig. 2 Representative MARS images: (**a**) Hand, 3D reconstruction of energy channel with positioning straps. (**b**) Hand, 3D reconstruction of bone. (**c**) Elbow, coronal slice from energy channel. (**d**) Wrist, sagittal slice from calcium (white), lipid (yellow) and water (red) channels. (**e**) Wrist, 3D reconstruction of calcium, lipid and water channels. (**f**) Wrist, 3D reconstruction of calcium, lipid and water channels

The MARS Extremity 5 × 120 reconstruction engine uses the photon counts to reconstruct 3D images, also called 3D maps or reconstructed volumes, of both the energy-dependent attenuation values and the material contents of each voxel. The images are presented to clinicians in the same form as conventional CT images in axial, coronal, sagittal and oblique planes. In addition, quantitative material information can be obtained and analysed using proprietary software for specified regions of interest using the spectral or multi-energy scans that have been described [23, 24].

Figure 2 shows a selection of scans obtained using the MARS Extremity 5 × 120 imaging system.

3 Applications of MARS Spectral Photon-Counting CT

3.1 Acute Fractures

As part of the world's first clinical trial of MARS spectral CT, the MARS Extremity 5 × 120 imaging system is currently being used to image patients with acute fractures of the upper limb. The imaging is taking place at a local radiology practice in an acute care clinic in the community. In its early stages, the trial has focused on wrist fractures, for example, scaphoid fractures resulting from a fall on an outstretched hand (FOOSH).

The scaphoid is one of eight carpal bones in the wrist and accounts for 2–7% of all fractures and 82–89% of carpal fractures [25]. Commonly occurring in young active adults, the location and degree of displacement of a scaphoid fracture play a key role in patient treatment and recovery [26]. However, up to 25% of suspected scaphoid fractures are radiographically occult and therefore cannot be confirmed on initial and sometimes on follow-up plain X-rays [27, 28]. Occult fractures are especially common in the wrist due to the complex three-dimensional relationship between the carpal bones and the superposition of anatomical structures on plain radiographs [29]. A delayed fracture diagnosis is associated with an increase in the incidence of avascular necrosis, non-union and osteoarthritis, an increased operative risk and a loss in productivity due to cast immobilisation [9, 11, 30, 31].

The diagnostic pathway for occult scaphoid fractures typically involves plain radiography with the addition of cross-sectional imaging within 1–2 weeks, such as computed tomography (CT), CBCT or magnetic resonance imaging (MRI) [32]. The current imaging pathway requires multiple modalities to reach a confirmed diagnosis. SPCCT technology within compact (small footprint) scanners shows the potential to add value to this imaging pathway where and when the patient presents as the scanners can be deployed in community clinics. The goal is to achieve an early diagnosis and minimise immobilisations and repeated clinical imaging.

The ongoing trial has demonstrated that MARS spectral photon-counting CT can identify acute scaphoid fractures in the clinical practice setting. The first case in this series shows a scaphoid waist fracture that was radiographically occult on initial radiographs and equivocal on follow-up radiographs. The fracture is shown in the MARS images in Fig. 3.

For the second case in this series, the MARS scans show high-resolution multiplanar reconstructions of a scaphoid waist fracture, acquired at the time of the patient's first presentation to the community clinic after a sporting injury (Fig. 4).

The triquetrum is the second most commonly fractured carpal bone, and similarly, fractures resulting from a FOOSH can lead to long-term complications such as non-union and arthritis [33]. Triquetrum fractures are difficult to diagnose on plain radiographs which means that cross-sectional modalities such as CT or MRI form an important part of the diagnostic pathway [33]. For this reason, the role of SPCCT is being investigated for this application. To date, there have been two participants recruited into the MARS clinical trial with suspected triquetrum fractures, one of

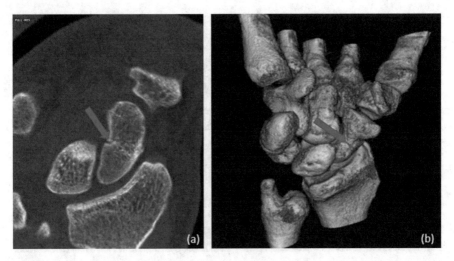

Fig. 3 MARS wrist scan showing the scaphoid waist fracture in the oblique (**a**) and 3D reconstruction (**b**)

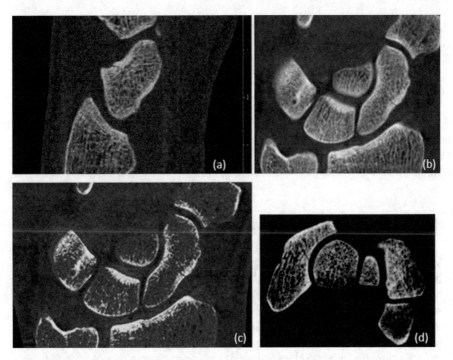

Fig. 4 MARS images show a scaphoid fracture in the sagittal oblique (**a**) and coronal plane (**b**) and calcium maps in the coronal (**c**) and axial plane (**d**)

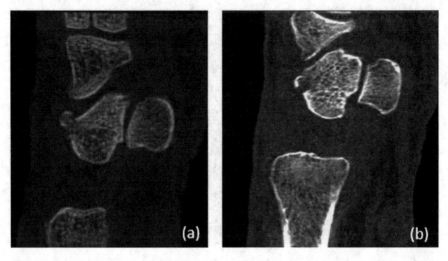

Fig. 5 MARS energy channel images in sagittal planes showing triquetrum fractures for two patients (**a, b**)

which was radiographically occult. Figure 5 shows that for each of these cases, the MARS imaging system was able to provide high-resolution cross-sectional imaging of the fracture when the patient presented to the community clinic.

The imaging of acute fractures has been primarily focused on the use of the dedicated upper limb scanner, the MARS Extremity 5 × 120. However, a prototype large bore MARS imaging system has also shown that the technology can be applied to lower limb imaging. Figure 6 demonstrates high-resolution 3D and 2D material maps of acute fractures of the fourth and fifth toe. The imaging was acquired less than 1 week from the date of injury.

3.2   Evaluation of Fracture Healing

The management of minimally displaced wrist fractures usually involves immobilisation in a well-fitted cast and, if necessary, plain radiographs to evaluate healing. Open, displaced or otherwise complex fractures typically warrant cross-sectional imaging such as CBCT to evaluate the position of fracture fragments and bone healing and to inform decision-making regarding surgical or conservative management [3]. Imaging needs to provide a distinction between a healing fracture and a fracture that is not healing and ideally provide information related to the status of the vasculature [34]. For scaphoid fractures, this is particularly important as the vascularity is often compromised when fractures are located at the proximal pole, resulting in bone loss or necrosis which further complicates patient recovery and long-term health outcomes [35, 36].

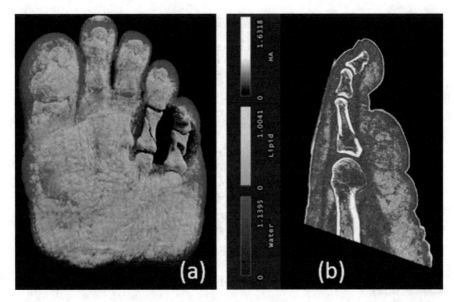

Fig. 6 (**a**) MARS 3D reformat showing fractures of the fourth and fifth proximal phalanx. (**b**) MARS calcium, lipid and water map in sagittal plane showing fracture of the fourth proximal phalanx

The MARS Extremity 5 × 120 imaging system is currently under investigation to evaluate whether the SPCCT technology is able to differentiate between bony union, fibrous union and non-union at fracture sites. The results, to date, confirm that MARS spectral CT is able to provide evidence of bone healing for scaphoid fractures and demonstrate the characteristic features of non-union.

The case shown in Fig. 7 demonstrates a healing scaphoid fracture, as evidenced by the central bone bridging across the fracture site in the energy channel images (a) and the calcium material maps (b). The calcium map provides the opportunity to quantify the calcium density and thereby further characterise the healing. Regions of interest drawn on the material map show the increase in calcium density at the location of the bone bridging (0.44 g/ml) compared to regions that are distal (0.20 g/ml) and proximal (0.24 g/ml) to the fracture site.

Similarly, for the clinical case in Fig. 8, the MARS scans show evidence of a healing scaphoid fracture where there is an observable increase in attenuation across the fracture site on the energy channel images.

MARS spectral CT has also been able to demonstrate the characteristic appearances of non-union in scaphoid fractures. Figure 9 shows the case of a patient with a comminuted scaphoid waist fracture who had ongoing pain 7 months after a fall on an outstretched hand. The sagittal MARS image shows a humpback deformity and sclerosis at the fracture margin, a sign that the fracture is not healing satisfactorily.

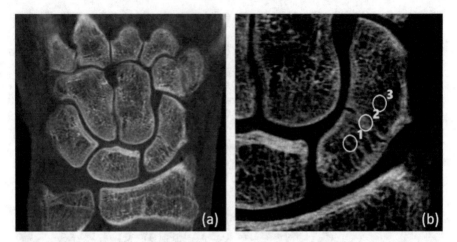

Fig. 7 MARS wrist imaging in a patient with scaphoid fracture. (**a**) Energy image shows bone bridging at the fracture site. (**b**) Calcium material map analysis shows the increase in bone mineralisation to 0.44 g/ml in ROI 2 compared to unaffected regions 0.24 g/ml in ROI 1 and 0.20 g/ml in ROI 3

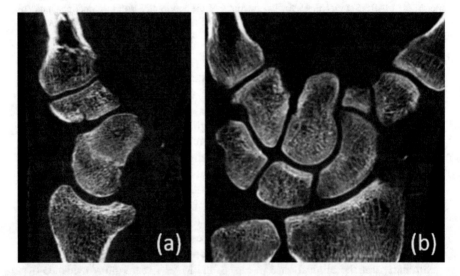

Fig. 8 Sagittal (**a**) and coronal (**b**) MARS imaging in a patient with a healing scaphoid fracture

3.3 Bone-Metal Interface Imaging

Post-operative imaging of metal implants is a challenge due to the susceptibility of CT to beam hardening artefact at the bone-metal interface. When the polychromatic X-ray beam interacts with material, the average energy of the beam is shifted to the higher energy side due to the preferential attenuation of lower energy photons,

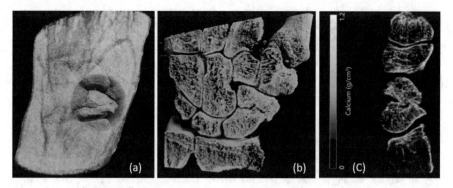

Fig. 9 (**a**, **b**) MARS 3D material reconstructions of the wrist with comminuted scaphoid waist fracture. (**c**) Sagittal calcium map demonstrates humpback deformity and sclerosis at the fracture margin

a process described as beam hardening [37]. This effect is particularly dominant in highly attenuating materials. A related effect is photon starvation: when the beam passes through highly attenuating objects, the detector is unable to receive an adequate number of photons [38]. The presence of high-attenuating objects in the path of X-ray beam causes nonlinear detector response resulting in dark and bright streaks known as streak artefacts [39]. The combined effects of beam hardening, streak artefact and photon starvation obscure the anatomy and pathology adjacent to metal implants, resulting in inconclusive diagnostic output [40].

The interface between metal and other structures is further limited by partial volume artefact which is the averaging of linear attenuation values when two differently attenuating objects are present in the same voxel. The partial volume artefact can be mitigated by employing high-resolution images and decreasing slice thickness. Hardware-based beam hardening correction techniques, such as pre-hardening of the beam, can be used to reduce these artefacts by employing filters; however, this approach produces poor signal-to-noise ratio (SNR) due to the elimination of soft X-rays which have better contrast differentiation for bone and for metal. Different beam hardening correction algorithms are typically employed to reduce the beam hardening but some create other artefacts as a by-product [41]. Generating virtual monochromatic energy images with dual energy CT technology can mitigate these artefacts; however, this does result in an increase in image noise at low energies [41, 42].

SPCCT has minimal metal artefact and high resolution and thus could help improve post-operative assessment of metal implants. Metallic hardware in the form of implants and prosthesis are very common in orthopaedic and dental applications to improve healing by providing stability and to restore functions by replacing defective parts of the bone [43]. Steel and titanium are common implant materials in the form of mesh, plate, screw or scaffolds due to their mechanical strength, corrosion resistance and biocompatibility [44]. Bone-metal interface imaging is important for the assessment of osseous integration which is determined by the

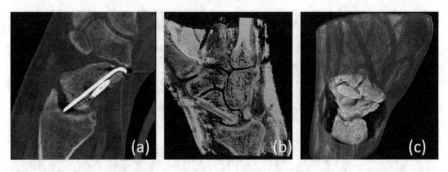

Fig. 10 MARS imaging in a patient with two screws fixation for scaphoid fracture. Artefact from metal hardware is minimal in all MARS images (**a**) Energy image, (**b**) HA and lipid material fused in 3D, (**c**) Water and HA material in 3D

tightness of the implant and integrity of the tissue in the vicinity of the implant [45, 46]. Significant costs for revision surgery can be reduced by earlier diagnosis of implant loosening or infection in the vicinity of the implant [47]. Assessment of the bone-metal interface can be difficult with other cross-sectional imaging modalities, including MRI and CBCT, and may be a particular opportunity for SPCCT.

Photon-counting detectors employed in MARS systems have the unique ability to capture images in multiple energy bins. Different implant materials when imaged with this technology have shown a reduction in artefacts in the higher energy bins of the spectrum [48]. The narrow energy bins suffer less from beam hardening and therefore can be utilised for the demonstration of the bone-metal interface. Simultaneously obtaining lower energy bin images means there is no compromise in terms of soft tissue differentiation [49].

Bridging bone and soft callus and associated mineralisation are markers of bone healing which are difficult to image in the vicinity of the metallic implants [50]. To monitor the dynamic process of bone healing, these markers have to be imaged with a high degree of accuracy. Through pre-clinical research, MARS has shown reliable results when compared with other technologies for the quantitative measurement of different parameters of bone healing, such as bone mineral density (BMD), bone mineral content (BMC) and the newly formed bone near metallic implants [49]. These pre-clinical research findings are currently being tested in the clinical environment. A series of post-surgical participants have been scanned with the MARS Extremity 5 × 120 imaging system as part of the ongoing clinical trial.

The case presented in Fig. 10 is an example of the high-resolution energy images that can be achieved in the presence of metal implants. The metal wire and screw are in close proximity and can be visualised as two separate structures with limited beam hardening artefact at the bone-metal interface.

The second clinical trial case shows high-resolution MARS images of a screw fixation of the lunate with minimal metal artefact at the bone-metal interface (Fig. 11).

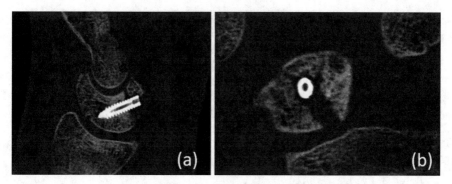

Fig. 11 MARS imaging in a patient with a screw fixation of the lunate. Streak artefact from metal hardware is minimal in the sagittal (**a**) and coronal (**b**) plane

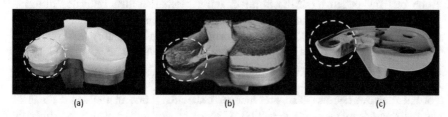

Fig. 12 Comparison between photograph and MARS images for the demonstration of arthroscopic implants defects. (**a**) Photograph of the implant from a patient with post-arthroplasty knee pain showing defects (inside dotted circle) in polyethylene components. (**b**) MARS 3D reconstruction showing defects similar to the photograph. (**c**) Demonstration of the defect in the metallic component is achieved by digitally extracting the polyethylene component. (Image courtesy of Lau et al. [8])

The clinical studies to date have demonstrated the potential applications of MARS SPCCT for imaging the bone-metal interface for orthopaedic implants at fracture sites. The technology could also have future applications for imaging of arthroscopic implants, which is another challenge due to the requirements for simultaneous assessment of metallic and polyethylene components which determine implant success. Earlier diagnosis of damage of arthroscopic knee implants can prevent significant costs related to major surgery and may change patient management [8]. In the published pre-clinical ex vivo study, MARS CT has shown the ability to demonstrate damage in both soft tissue and metallic components of arthroscopic implants which have not been demonstrated accurately with other technologies including MRI, USG and X-ray [8]. The ability of MARS spectral CT to demonstrate the defect in both polyethylene and metallic components of knee arthroscopic implants is compared with the images in Fig. 12. A comparative study between MARS and cone beam CT for the evaluation of dental implants has also shown equivalent capability of MARS for the demonstration of the accessory canal and has superior ability for the imaging of implant and tooth interface [51].

Compared to energy-integrating CT systems intended for peripheral imaging, such as high-resolution peripheral quantitative CT and cone beam CT, MARS has shown improved ability to demonstrate the bone-metal interface due to less artefact from high-attenuating metal. In Figs. 10 and 11, the streak in the region of the implants is significantly lower in the MARS images than would be expected with other technologies. This ability to obtain both bone detail and soft tissue information should simplify the imaging of patients with metallic implants and prosthesis and therefore result in better treatment outcome.

3.4 Bone Marrow Oedema Measurements

Bone marrow oedema (BME) is an imaging biomarker that indicates different pathophysiological conditions of the bone [52]. There is an accumulation of fluid and other fluid-like components (haemorrhage and inflammatory infiltrate) into marrow space as a result of trauma, inflammation, neoplasm and other pathologies. This collection of water-like components replaces the lipid-like component in marrow space and creates a small increase in X-ray attenuation with respect to normal marrow regions. The presence of trabecular bone in the same voxel as the marrow complicates the detection of small proportional changes of either fat or water and provides a challenge for the detection of this physiological process [6]. Similarly, the detection of marrow oedema is more challenging when a higher proportion of red marrow is present in regions of interest such as the spine and when the accumulation of fluid-like material is small [53–55].

Detection of bone marrow oedema as an imaging marker is important for the earlier detection and differential diagnosis of different bone-related pathologies, and to monitor the progression of disease conditions [56]. Occult fractures related to impaction injuries of trabecular bone are identified with the help of bone marrow oedema [57, 58]. As an example, some carpal bone fractures, particularly scaphoid bones, are not visible in the acute phase of injury, and these fractures are only diagnosed after assessment of changes in marrow signal [1, 59]. Marrow oedema can also be seen in degenerative, ischemic and metastatic conditions of the bone as an indication of disease severity and progression of disease. The presence and extent of oedema in bone due to these conditions can also be representative of the severity of pain [60]. Therefore, bone marrow oedema can explain the various underlying causes before morphological changes are seen in the bone.

Small attenuation variations in marrow space are difficult to detect with conventional CT due to the obscuring of marrow with the high-attenuating signals of surrounding trabecular bone [6]. However, low-attenuating marrow cells and high-attenuating cancellous bone can produce individual spectral signatures when imaged with a spectral CT imaging system [61], and subtle attenuation variation in the marrow region can be seen when the trabecular bone component is separated from marrow using material identification techniques [62, 63]. Dual energy CT can perform material separation and has shown promising results with regard to the demonstration of bone marrow oedema using virtual non-calcium (VNCa) techniques [53, 58] and three material decomposition algorithms [64].

MARS technology can produce images in at least five pre-defined energy bins and, using multiple energy bins, can produce material maps. The energy thresholds in this system can be optimised to provide more weighting in lower energy bins where better separation between lipid and water can be achieved [49].

To detect bone marrow oedema with MARS technology, three material maps are created using proprietary algorithms which are calibrated using lipid, water and hydroxyapatite phantoms. The oedematous region is the area with an increase in water mass density and decrease in lipid mass density. This application was tested in a clinical trial case, where a fractured toe was scanned in the acute phase using a large bore MARS spectral CT scanner.

The MARS energy channel images in Fig. 13 show a normal third proximal phalanx and fractured fourth and fifth proximal phalanges. When the lipid map was fused with the energy channel, the lipid density could be observed in the medullary cavity of each of the phalanges. There is an obvious and measurable reduction in lipid density in the fractured phalanges compared to the normal phalanx. Conversely, the water maps showed an obvious increase in water density in the medullary cavity of the fractured phalanges compared to the normal phalanx as expected. Regions of interest can be drawn to quantify the changes in material density which may be useful for monitoring changes in oedema over time.

It is expected that MARS spectral CT imaging systems will have applications for the identification of oedema, for the early identification of occult fractures and for the monitoring of oedematous changes over time.

3.5 Bone Quality Measures: Bone Structure and Material Maps

Structurally, bone consists of the outer compact or cortical bone and the inner cancellous or trabecular bone. Based on X-ray attenuation, bone is comprised of three major material compositions (bone mineral equivalent, fat and soft tissue equivalent). Structural alteration and proportional variation in intrinsic material occur in various pathological conditions that result in abnormal bone [65, 66].

Hydroxyapatite is the major inorganic component of bone comprising almost 70% by mass proportion and almost 50% by volume proportion. The mass of an inorganic component determines the rigidity of bone. If the mass proportion of minerals in the bone decreases, the bones become thin and fragile, such as in the condition of osteoporosis [67]. Osteoporosis is more common in the elderly population, especially postmenopausal women, patients receiving corticosteroid treatment and patients with vitamin and mineral deficiencies [68, 69].

The direct structural information of the inner compartment (trabecular) bone is important for fracture risk prediction. The inner compartment of bone has a high

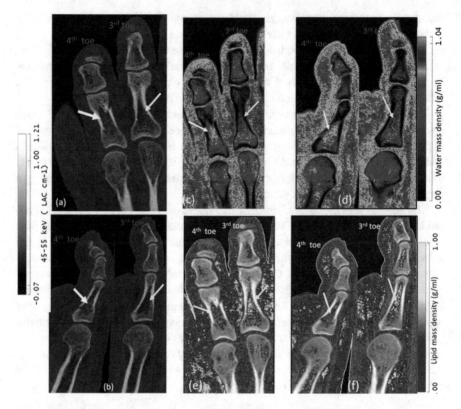

Fig. 13 Quantitative assessment of bone marrow oedema using MARS images. (**a, b**) Energy images in the coronal and sagittal plane showing fracture in the proximal phalanx of the fourth toe. (**c, d**) Water map showing increased water concentration in bone marrow region near fracture s/o of bone marrow oedema. The corresponding bone marrow region shows decrease in lipid concentration (**e, f**), as compared to normal marrow region in the proximal phalanx of the third toe. The marrow of the fracture region is indicated with white arrow and the corresponding normal marrow in the third toe is indicated with yellow arrows

turnover rate and osteoporosis-related bone loss is more severe in this compartment; therefore, this region is prone to osteoporosis-related fractures [70]. Quantitative and structural assessment of trabecular bone can provide information regarding dimensions, shape and mineral content to support the prediction of fracture risk. To obtain these structural details, high-resolution images are required with resolution close to the order of trabecular size.

Currently in clinical practice, bone quality measurement is achieved using DEXA which may be combined with clinical fracture risk assessment tools such as FRAX, and trabecular bone scoring. However, as discussed in the background to this chapter, DEXA is not able to provide 3D multiplanar imaging of cortical and trabecular bone structure and is inaccurate in the presence of metal implants. A novel MRI technique, sweep imaging with Fourier transform (SWIFT), is currently being explored in pre-clinical research as it has been shown to measure bound water in cortical bone which is associated with cortical bone formation [71]. CT techniques, such as high-resolution peripheral quantitative CT (HR-PQCT) and peripheral cone beam CT, can provide three-dimensional bone detail; however, these technologies are limited to peripheral regions only. Furthermore, they cannot measure the subtle variation in the mineral component as well as variation in water and lipid components in the bone marrow. The bone marrow is another major factor in determining bone strength [72] and is classified according to the hematopoietic marrow (red marrow) and the fat-like yellow marrow [73]. Therefore, quantitative assessment of maps of all three components can better explain bone health in various situations.

The MARS imaging system provides simultaneous high-resolution images of bone for structural assessment and material-specific maps for the assessment of material composition of bone. The voxel size of 90 μm is closer to the required resolution for trabecular bone detail compared to other modalities, and the small voxel sizes mean less partial volume averaging. This should result in better material identification and quantification which translates to more accurate bone health estimation.

Lipid-, water- and bone mineral-specific maps are produced using MARS material identification and quantification algorithms [74]. The mass attenuation values in different energy bins are obtained using the linear attenuation and density of materials [74]. The linear attenuation value of calcium in bone decreases dramatically at a higher energy compared to lipid and water in bone. This property of calcium, along with its unique density values, helps to generate calcium-specific maps during the process of material identification. Serial phantom data, with hydroxyapatite rods of known concentrations and vials of lipid and water, are obtained and compared with reference values for the assessment of reproducibility of the system. MARS technology has shown average measurement errors and root mean square errors less than 10 mg/cm^3 for bone mineral estimation as demonstrated in Fig. 14a.

In pre-clinical research, the comparison of MARS results using DEXA and HRpQCT have shown that MARS is congruent with other existing technologies used for bone health estimation [75]. Clinical research is now underway into the use of MARS imaging systems for bone health assessment. The wrist is one of the most common osteoporotic fracture sites. The research team are analysing wrist images that were obtained at point-of-care for other purposes, such as suspected carpal bone fractures. Measurements are taken from a 9 mm region of the distal radius with the radio-ulnar joint as a reference point (Fig. 14a). This reflects the ultra-distal site that is currently used in clinical practice for DEXA. The three material maps (water, lipid and HA) produced from the material identification algorithms are demonstrated in Fig. 14c–e, respectively. The hydroxyapatite map in (e) can be further segmented into cortical and trabecular compartments as shown in Fig. 14f, g, respectively. Different bone mineral parameters have been computed using the hydroxyapatite material maps generated by the MARS imaging system. The total bone mineral density (TBMD) gives the average bone mineral density value of both cortical and trabecular compartments of bone. The cortical and trabecular compartments can be

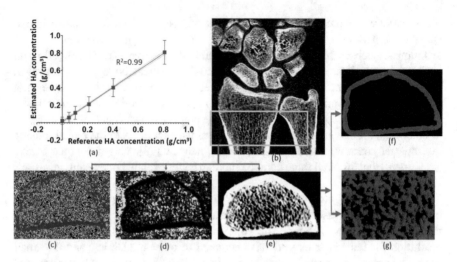

Fig. 14 Simultaneous measurement of bone mineral density, microstructures and material maps of human distal radius with MARS spectral CT. (**a**) Linearity graph showing linear correlation between reference and estimated values for the HA rods in the calibration phantom. (**b**) Coronal image of the wrist showing slice selection regions for the measurement of bone mineral density in ultra-distal radius. (**c–e**) axial image of distal radius showing water, lipid and HA maps, respectively. The hydroxyapatite map can be further segmented into cortical (**f**) and trabecular bone (**g**) to estimate bone mineral density in each compartment

segmented using thresholding techniques to obtain separate cortical bone mineral density (CBMD) and trabecular bone mineral density (TBMD) measurements [17]. Measurements are also obtained for trabecular thickness (TT) and trabecular spacing (TS). In pathological conditions, the trabecular bone volumes can deteriorate and the shape changes from plate-like structures to rod-like structures [76]. This change in trabecular shape and volume creates inhomogeneity in the trabecular region. The measurements obtained from the MARS imaging system should capture these pathological changes. Measurement of the above-mentioned densitometric and geometric parameters provides more information regarding bone health. However, further studies are required regarding the scaling up of MARS extremity scanner to a body scanner for correlating wrist measurements with other prominent fracture sites such as the hip and spine.

In addition, MARS provides additional material maps for lipid and water density. The quantitative assessment of the proportional variation of bone marrow adipose tissue has been shown to be a better prediction of bone strength in various pathophysiological conditions like diabetes, obesity and anorexia nervosa [77–79]. Similarly, the increase in bone marrow water component, such as in the case of bone marrow oedema, can lead to weak and fragile bone [80]. The utility of material maps for the detection of bone marrow edema was discussed in the previous section.

Opportunistic bone health assessment allows simultaneous bone imaging and bone health assessment without further radiation dose. It is expected that significant numbers of patients who are referred for upper extremity examinations could benefit from this approach as it removes barriers such as cost, access and convenience. The overall goal is for earlier screening to decrease the number of osteoporosis-related fractures, thus reducing healthcare costs and improving the patient's quality of life.

4 Summary

MARS Bioimaging Ltd. and its various partners have shown the clinical utility of the MARS Extremity 5 × 120 as a point-of-care, high spatial resolution and high-energy resolution spectral CT imaging system. Clinical trials have demonstrated its applications for acute fractures, the evaluation of fracture healing and the assessment of osseous integration at the bone-metal interface for fracture fixations and joint replacements. Material maps provide the opportunity to evaluate bone marrow oedema for the detection of occult fractures, for monitoring of disease and for the opportunistic evaluation of bone health. In the future, the design of point-of-care MARS scanners for lower extremity and head and neck imaging will further extend these clinical applications.

Acknowledgements This project was funded by the Ministry of Business, Innovation and Employment (MBIE), New Zealand, under contract number UOCX1404, by MARS Bioimaging Ltd. and the Ministry of Education through the MedTech CoRE. The authors would like to acknowledge the Medipix2, Medipix3 and Medipix4 collaborations. Also, they would like to take this opportunity to acknowledge the generous support of the MARS Collaboration.

MARS Collaboration Sikiru A. Adebileje[a,e]; Steven D. Alexander[a]; Maya R. Amma[a]; Marzieh Anjomrouz[a]; Fatemeh Asghariomabad[a]; Ali Atharifard[a]; James Atlas[b]; Stephen T. Bell[a]; Anthony P. H. Butler[a,b,c,d,e]; Philip H. Butler[a,b,c,d,e]; Pierre Carbonez[c,d]; Claire Chambers[a]; Krishna M. Chapagain[a,c]; Alexander I. Chernoglazov[a,e]; Jennifer A. Clark[a,c,f]; Frances Colgan[c]; Jonathan S. Crighton[c]; Shishir Dahal[c,i,j]; Jérôme Damet[c,d,k]; Theodorus Dapamede[c]; Niels J. A. de Ruiter[a,b,c,e]; Devyani Dixit[a,b]; Robert M. N. Doesburg[a]; Karen Dombroski[a]; Neryda Duncan[a]; Steven P. Gieseg[b,c,d]; Anish Gopinathan[a]; Brian P. Goulter[a]; Joseph L. Healy[a]; Luke Holmes[l]; Kevin Jonker[a,c]; Tracy Kirkbride[f]; Chiara Lowe[a]; V. B. H. Mandalika[a,e]; Aysouda Matanaghi[a]; Mahdieh Moghiseh[a,c]; Manoj Paladugu[b]; David Palmer[g]; Raj K. Panta[a,d]; Peter Renaud[b,c]; Yann Sayous[b]; Nanette Schleich[h]; Emily Searle[b]; Jereena S. Sheeja[a]; Aaron Smith[b]; Leza Vanden Broeke[a]; V. S. Vivek[a]; E. Peter Walker[c], Manoj Wijesooriya[a]; W. Ross Younger[a]

[a]MARS Bioimaging Limited, Christchurch, New Zealand
[b]University of Canterbury, Christchurch, New Zealand
[c]University of Otago Christchurch, Christchurch, New Zealand
[d]European Organisation for Nuclear Research (CERN), Geneva, Switzerland
[e]Human Interface Technology Laboratory New Zealand, University of Canterbury, Christchurch, New Zealand
[f]Ara Institute of Canterbury, Christchurch, New Zealand
[g]Lincoln University, Lincoln, New Zealand
[h]University of Otago Wellington, Wellington, New Zealand
[i]Ministry of Health, Kathmandu, Nepal
[j]National Academy of Medical Sciences, Kathmandu, Nepal
[k]Institute of Radiation Physics, Lausanne University Hospital, Lausanne, Switzerland
[l]Canterbury District Health Board, Christchurch, New Zealand

References

1. Geijer M. Diagnosis of scaphoid fracture: optimal imaging techniques. Rep Med Imaging. 2013;6:57–69.
2. Bishop JA, et al. Assessment of compromised fracture healing. JAAOS: J Am Acad Orthop Surg. 2012;20(5):273–82.
3. Colville J, et al. Evaluating cone-beam CT in the diagnosis of suspected scaphoid fractures in the emergency department: preliminary findings. Clin Imaging. 2022;83:65–71.
4. Snaith B, et al. Evaluating the potential for cone beam CT to improve the suspected scaphoid fracture pathway: InSPECTED-A single-centre feasibility study. J Med Imaging Radiat Sci. 2022;53(1):35–40.
5. Demehri S, et al. Assessment of image quality in soft tissue and bone visualization tasks for a dedicated extremity cone-beam CT system. Eur Radiol. 2015;25(6):1742–51.
6. Pache G, et al. Dual-energy CT virtual noncalcium technique: detecting posttraumatic bone marrow lesions—feasibility study. Radiology. 2010;256(2):617–24.
7. Palmer WE, Simeone FJ. Can dual-energy CT challenge MR imaging in the diagnosis of focal infiltrative bone marrow lesions? Radiol Soc N Am. 2018;286(1):214–6.
8. Lau LCM, et al. Multi-energy spectral photon-counting computed tomography (MARS) for detection of arthroplasty implant failure. Sci Rep. 2021;11(1):1–6.
9. Mallee WH, et al. Computed tomography versus magnetic resonance imaging versus bone scintigraphy for clinically suspected scaphoid fractures in patients with negative plain radiographs. Cochrane Database Syst Rev. 2015;6.
10. Memarsadeghi M, et al. Occult scaphoid fractures: comparison of multidetector CT and MR imaging—initial experience. Radiology. 2006;240(1):169–76.
11. Ahn JM, El-Khoury GY. Occult fractures of extremities. Radiol Clin N Am. 2007;45(3):561–79. ix
12. Breitenseher MJ, et al. Radiographically occult scaphoid fractures: value of MR imaging in detection. Radiology. 1997;203(1):245–50.
13. Jain R, et al. Early scaphoid fractures are better diagnosed with ultrasonography than X-rays: a prospective study over 114 patients. Chin J Traumatol. 2018;21(04):206–10.
14. van Hamersvelt RW, et al. Accuracy of bone mineral density quantification using dual-layer spectral detector CT: a phantom study. Eur Radiol. 2017;27(10):4351–9.
15. Lindgren E, Rosengren BE, Karlsson MK. Does peak bone mass correlate with peak bone strength? Cross-sectional normative dual energy X-ray absorptiometry data in 1052 men aged 18–28 years. BMC Musculoskelet Disord. 2019;20(1):1–10.
16. Link TM, Kazakia G. Update on imaging-based measurement of bone mineral density and quality. Curr Rheumatol Rep. 2020;22(5):1–11.
17. Matanaghi A, et al. Semi-automatic quantitative assessment of site-specific bone health using spectral photon counting CT. J Nucl Med. 2019;60(supplement 1):1297.
18. Roessl E, Proksa R. K-edge imaging in x-ray computed tomography using multi-bin photon counting detectors. Phys Med Biol. 2007;52(15):4679.
19. Jones A, Hintenlang D, Bolch W. Tissue-equivalent materials for construction of tomographic dosimetry phantoms in pediatric radiology. Med Phys. 2003;30(8):2072–81.
20. Coursey CA, et al. Dual-energy multidetector CT: how does it work, what can it tell us, and when can we use it in abdominopelvic imaging? Radiographics. 2010;30(4):1037–55.
21. Ballabriga R, et al. The Medipix3RX: a high resolution, zero dead-time pixel detector readout chip allowing spectroscopic imaging. J Instrum. 2013;8(02):C02016.
22. Aamir R, et al. MARS spectral molecular imaging of lamb tissue: data collection and image analysis. J Instrum. 2014;9(02):P02005.
23. Panta RK, et al. First human imaging with MARS photon-counting CT. In: 2018 IEEE nuclear science symposium and medical imaging conference proceedings (NSS/MIC). IEEE; 2018.
24. Willemink MJ, et al. Photon-counting CT: technical principles and clinical prospects. Radiology. 2018;289(2):293–312.

25. Rhemrev SJ, et al. Current methods of diagnosis and treatment of scaphoid fractures. Int J Emerg Med. 2011;4(1):1–8.
26. Gilley E, et al. Importance of computed tomography in determining displacement in scaphoid fractures. J Wrist Surg. 2018;7(01):038–42.
27. Dorsay TA, Major NM, Helms CA. Cost-effectiveness of immediate MR imaging versus traditional follow-up for revealing radiographically occult scaphoid fractures. Am J Roentgenol. 2001;177(6):1257–63.
28. Jenkins PJ, et al. A comparative analysis of the accuracy, diagnostic uncertainty and cost of imaging modalities in suspected scaphoid fractures. Injury. 2008;39(7):768–74.
29. Kaewlai R, et al. Multidetector CT of carpal injuries: anatomy, fractures, and fracture-dislocations. Radiographics. 2008;28(6):1771–84.
30. Gelberman RH, Wolock B, Siegel DB. Current concepts review: fractures and non-unions of the carpal scaphoid. J Bone Joint Surg Am. 1989;71(10):1560–5.
31. Gupta V, Rijal L, Jawed A. Managing scaphoid fractures. How we do it? J Clin Orthop Trauma. 2013;4(1):3–10.
32. Bhat M, et al. MRI and plain radiography in the assessment of displaced fractures of the waist of the carpal scaphoid. J Bone Joint Surg Br Vol. 2004;86(5):705–13.
33. Guo RC, Cardenas JM, Wu CH. Triquetral Fractures Overview. Curr Rev Musculoskelet Med. 2021;14(2):101–6.
34. Donati OF, et al. Is dynamic gadolinium enhancement needed in MR imaging for the preoperative assessment of scaphoidal viability in patients with scaphoid nonunion? Radiology. 2011;260(3):808–16.
35. Soufi M, See A, Hassan S. Scaphoid fractures and non-union: a review of current evidence. Orthop Trauma. 2021;35(4):198–207.
36. Amrami KK, Frick MA, Matsumoto JM. Imaging for acute and chronic scaphoid fractures. Hand Clin. 2019;35(3):241–57.
37. McKetty MH. The AAPM/RSNA physics tutorial for residents. X-ray attenuation. Radiographics. 1998;18(1):151–63.
38. Mori I, et al. Photon starvation artifacts of X-ray CT: their true cause and a solution. Radiol Phys Technol. 2013;6(1):130–41.
39. Mouton A, et al. An experimental survey of metal artefact reduction in computed tomography. J Xray Sci Technol. 2013;21(2):193–226.
40. Hunter AK, McDavid W. Characterization and correction of cupping effect artefacts in cone beam CT. Dentomaxillofac Radiol. 2012;41(3):217–23.
41. Wu R, et al. Quantitative comparison of virtual monochromatic images of dual energy computed tomography systems: beam hardening artifact correction and variance in computed tomography numbers: a phantom study. J Comput Assist Tomogr. 2018;42(4):648–54.
42. Kalisz K, et al. Noise characteristics of virtual monoenergetic images from a novel detector-based spectral CT scanner. Eur J Radiol. 2018;98:118–25.
43. Pawelec K, Planell JA. Bone repair biomaterials: regeneration and clinical applications. Elsevier Ltd; 2018.
44. Prasad K, et al. Metallic biomaterials: current challenges and opportunities. Materials. 2017;10(8):884.
45. Wirth AJ, et al. Implant stability is affected by local bone microstructural quality. Bone. 2011;49(3):473–8.
46. Shah FA, et al. Long-term osseointegration of 3D printed CoCr constructs with an interconnected open-pore architecture prepared by electron beam melting. Acta Biomater. 2016;36:296–309.
47. Cyteval C, Bourdon A. Imaging orthopedic implant infections. Diagn Interv Imaging. 2012;93(6):547–57.
48. Rajendran K, et al. Reducing beam hardening effects and metal artefacts in spectral CT using Medipix3RX. J Instrum. 2014;9(03):P03015.
49. Rajeswari Amma M. Study of bone-metal interface in orthopaedic application using spectral CT. University of Otago; 2020.

50. Fisher JS, et al. Radiologic evaluation of fracture healing. Skelet Radiol. 2019;48(3):349–61.
51. Broeke LV, et al. Feasibility of photon-counting spectral CT in dental applications—a comparative qualitative analysis. BDJ Open. 2021;7(1):1–8.
52. Manara M, Varenna M. A clinical overview of bone marrow edema. Reumatismo. 2014;66:184–96.
53. Diekhoff T, et al. Dual-energy CT virtual non-calcium technique for detection of bone marrow edema in patients with vertebral fractures: a prospective feasibility study on a single-source volume CT scanner. Eur J Radiol. 2017;87:59–65.
54. Kosmala A, et al. Multiple myeloma and dual-energy CT: diagnostic accuracy of virtual noncalcium technique for detection of bone marrow infiltration of the spine and pelvis. Radiology. 2018;286(1):205–13.
55. Gosangi B, et al. Bone marrow edema at dual-energy CT: a game changer in the emergency department. Radiographics. 2020;40(3):859–74.
56. Suh CH, et al. Diagnostic performance of dual-energy CT for the detection of bone marrow oedema: a systematic review and meta-analysis. Eur Radiol. 2018;28(10):4182–94.
57. Ali IT, et al. Clinical utility of dual-energy CT analysis of bone marrow edema in acute wrist fractures. Am J Roentgenol. 2018;210(4):842–7.
58. Dareez NM, et al. Scaphoid fracture: bone marrow edema detected with dual-energy CT virtual non-calcium images and confirmed with MRI. Skelet Radiol. 2017;46(12):1753–6.
59. Taljanovic MS, et al. Imaging and treatment of scaphoid fractures and their complications. In: Seminars in musculoskeletal radiology. Thieme Medical Publishers; 2012.
60. Collins JA, et al. Bone marrow edema: chronic bone marrow lesions of the knee and the association with osteoarthritis. Bull NYU Hosp Jt Dis. 2016;74(1):24.
61. Patino M, et al. Material separation using dual-energy CT: current and emerging applications. Radiographics. 2016;36(4):1087–105.
62. Bierry G, et al. Dual-energy CT in vertebral compression fractures: performance of visual and quantitative analysis for bone marrow edema demonstration with comparison to MRI. Skelet Radiol. 2014;43(4):485–92.
63. Akisato K, et al. Dual-energy CT of material decomposition analysis for detection with bone marrow edema in patients with vertebral compression fractures. Acad Radiol. 2020;27(2):227–32.
64. Schwaiger BJ, et al. Three-material decomposition with dual-layer spectral CT compared to MRI for the detection of bone marrow edema in patients with acute vertebral fractures. Skelet Radiol. 2018;47(11):1533–40.
65. Mikolajewicz N, et al. HR-pQCT measures of bone microarchitecture predict fracture: systematic review and meta-analysis. J Bone Miner Res. 2020;35(3):446–59.
66. Sornay-Rendu E, et al. Bone microarchitecture assessed by HR-pQCT as predictor of fracture risk in postmenopausal women: the OFELY study. J Bone Miner Res. 2017;32(6):1243–51.
67. Seeman E. Reduced bone formation and increased bone resorption: rational targets for the treatment of osteoporosis. Osteoporos Int. 2003;14(3):2–8.
68. Nieves J, et al. High prevalence of vitamin D deficiency and reduced bone mass in multiple sclerosis. Neurology. 1994;44(9):1687.
69. Nuti R, et al. Guidelines for the management of osteoporosis and fragility fractures. Intern Emerg Med. 2019;14(1):85–102.
70. Fuggle NR, et al. Fracture prediction, imaging and screening in osteoporosis. Nat Rev Endocrinol. 2019;15(9):535–47.
71. Sotozono Y, et al. Sweep imaging with Fourier transform as a tool with MRI for evaluating the effect of teriparatide on cortical bone formation in an ovariectomized rat model. BMC Musculoskelet Disord. 2022;23(1):1–10.
72. Hawkes C, Mostoufi-Moab S. Fat-bone interaction within the bone marrow milieu: impact on hematopoiesis and systemic energy metabolism. Bone. 2019;119:57–64.
73. Ambrosi TH, Schulz TJ. The emerging role of bone marrow adipose tissue in bone health and dysfunction. J Mol Med. 2017;95(12):1291–301.

74. Bateman CJ, et al. MARS-MD: rejection based image domain material decomposition. J Instrum. 2018;13(05):P05020.
75. Ramyar M, et al. Establishing a method to measure bone structure using spectral CT. In Medical imaging 2017: physics of medical imaging. International Society for Optics and Photonics; 2017.
76. Wang J, et al. Trabecular plates and rods determine elastic modulus and yield strength of human trabecular bone. Bone. 2015;72:71–80.
77. Verma S, et al. Adipocytic proportion of bone marrow is inversely related to bone formation in osteoporosis. J Clin Pathol. 2002;55(9):693–8.
78. Patsch JM, et al. Bone marrow fat composition as a novel imaging biomarker in post-menopausal women with prevalent fragility fractures. J Bone Miner Res. 2013;28(8):1721–8.
79. Hassan EB, et al. Bone marrow adipose tissue quantification by imaging. Curr Osteoporos Rep. 2019;17(6):416–28.
80. Zanetti M, et al. Bone marrow edema pattern in osteoarthritic knees: correlation between MR imaging and histologic findings. Radiology. 2000;215(3):835–40.

MARS for Molecular Imaging and Preclinical Studies

Mahdieh Moghiseh, Jennifer A. Clark, Maya R. Amma,
Krishna M. Chapagain, Devyani Dixit, Chiara Lowe, Aysouda Matanaghi,
Emily Searle, Yann Sayous, Dhiraj Kumar, and Anthony P. H. Butler

1 Introduction

The previous chapter described the first clinical MARS imaging system: a dedicated upper extremity SPCCT scanner capable of high-resolution orthopaedic imaging. The clinical MARS imaging system is a product of 15 years of preclinical research and development of a small-bore human translatable SPCCT system. The current

M. Moghiseh (✉) · J. A. Clark · K. M. Chapagain
MARS Bioimaging Limited, Christchurch, New Zealand

Department of Radiology, University of Otago Christchurch, Christchurch, New Zealand
e-mail: mahdieh.moghiseh@marsbioimaging.com

M. R. Amma · C. Lowe · A. Matanaghi
MARS Bioimaging Limited, Christchurch, New Zealand

D. Dixit · E. Searle · Y. Sayous
MARS Bioimaging Limited, Christchurch, New Zealand

Free Radical Biochemistry Laboratory, School of Biological Sciences, University of Canterbury, Christchurch, New Zealand

D. Kumar
University of Minnesota, Minneapolis, MN, USA

A. P. H. Butler
MARS Bioimaging Limited, Christchurch, New Zealand

Department of Radiology, University of Otago Christchurch, Christchurch, New Zealand

Free Radical Biochemistry Laboratory, School of Biological Sciences, University of Canterbury, Christchurch, New Zealand

European Organisation for Nuclear Research (CERN), Geneva, Switzerland

Human Interface Technology Laboratory New Zealand, University of Canterbury, Christchurch, New Zealand

© The Author(s), under exclusive license to Springer Nature Switzerland AG 2023
S. Hsieh, K. (Kris) Iniewski (eds.), *Photon Counting Computed Tomography*,
https://doi.org/10.1007/978-3-031-26062-9_4

chapter will discuss the valuable role SPCCT technology could play in molecular imaging applications, including for infectious diseases, cancer, atherosclerosis, bone health, cartilage health and joint health. The gold standard imaging for each of the clinical applications will be reviewed followed by discussion of the applications of MARS spectral photon-counting CT for molecular imaging.

2 Background

Molecular imaging in medicine is used to visualize, characterize and quantify pathophysiological processes in vivo. Currently, cross-sectional imaging in clinical practice is performed using computed tomography (CT), magnetic resonance imaging (MRI), ultrasound (US), positron emission tomography (PET) and single-photon emission computed tomography (SPECT). Each of the clinical techniques has unique advantages and limitations with respect to molecular imaging.

CT is a reliable imaging technique that provides excellent anatomical information and has a robust spatial resolution. However, the lack of high contrast sensitivity makes CT inadequate for molecular imaging. As compared to CT, MRI provides better soft tissue contrast and does not require radiation, but the spatial resolution is inferior to CT. Nevertheless, the main obstacle to employing MRI for molecular imaging is limited contrast sensitivity. Ultrasound is another morphological imaging tool that is widely used in clinical practice. The principal weakness of ultrasound is the depth of penetration, which limits the applications and makes whole-body imaging impossible. Contrast agents and molecular probes can pave the way to enable molecular imaging using CT, MRI and US [3–7]. However, the development of suitable contrast agents and molecular probes presents an extremely difficult challenge [4, 5, 8–10]. Currently, there are limited nano-particles contrast agents that have been approved by major regulatory to serve as molecular probes for use with MRI [1, 2].

PET and SPECT have been the most common molecular imaging techniques used in clinical settings for the last decade [11, 12]. Although PET and SPECT have excellent sensitivity (ng/mL) and can provide functional information, they require radioactive pharmaceutical agents and thus are associated with an increased radiation dose to the patient [13]. Furthermore, PET suffers from poor spatial resolution and lacks anatomical information [13, 14]. In order to improve spatial resolution and have anatomical localization of the tracer activity within the body, PET and SPECT are combined with other imaging modalities, such as CT. However, studies have shown that PET/CT doses are higher than those of other imaging modalities [15]. The production of the short-lived radioisotopes in a cyclotron and the transport to the clinical centre adds a significant cost to the PET imaging. Another challenge with PET is that the most common PET pharmaceutical (FDG) measures only one molecular process, so it is hard to directly quantify other biological processes besides metabolic activity. Also, PET and SPECT are lengthy procedures [16, 17], with patients needing to designate around 2–3 hours for the entire appointment, including time for the tracer to travel throughout the body after injection and the scan itself. In addition to all these limitations, widespread use of PET is limited and is typically restricted to oncological and some orthopaedic applications.

In addition to clinical applications, molecular imaging plays a significant role in preclinical research. It has been considered to be one of the most important techniques for drug discovery and developing new agents to facilitate targeting imaging and personalized treatments. The same imaging modalities that are used in the clinical setting can also be used for preclinical examinations. However, as well as all the limitations listed above, these imaging modalities are very costly and not accessible to every lab for scanning small animals and specimens.

Micro-CT is the imaging technique that is often used in preclinical animal studies to assess disease burden and drug efficacy. Micro-CT provides very high spatial resolution (100 nm) and 3D information that cannot be obtained by any other non-destructive method [18]. Diseased tissue and contrast material are identified through abnormal x-ray densities. The disadvantage of micro-CT is that preclinical applications cannot be directly translated to clinical human imaging due to micro-CT operating at a very high dose of ionizing radiation and technological differences between micro-CT and standard clinical CT systems. Scaling up micro-CT to human imaging would result in harmful exposure levels, therefore preventing direct translation of micro-CT findings to clinical CT. Preclinical imaging that uses the same hardware and software technologies as clinical imaging would result in a more rapid translation.

Spectral photon-counting CT (SPCCT) offers a low-cost molecular imaging option by using intrinsic biomarkers and introducing non-radioactive, high atomic number (high-Z) agents. Thus, SPCCT may open the door to a broader range of medical investigations and increase patient access to advanced diagnostics. The ability of SPCCT to identify calcifications and contrast agents and provide accurate soft tissue contrast in a single acquisition CT scan is an advantage this technology has over current functional molecular imaging modalities used in the clinic [19–23]. More importantly, multiple pharmaceuticals can be administered to the same subject simultaneously to investigate more than one molecular process in a single scan. SPCCT overcomes the limitations of PET and SPECT imaging, with high spatial resolution (down to 100 microns), low costs and radiation dose, is not labour intensive and does not require handling of radioactive contrast agents. Furthermore, some molecular processes provide enough intrinsic contrast to not need an administered agent. For example, while calcium and iron are close on the periodic table, the spectral nature of SPCCT enables the two atoms to be differentiated [24]. Using SPCCT, it is possible to quantify a molecular process by measuring functionalized high-Z pharmaceuticals (mg/mL) targeted to a molecular process [25]. Moreover, SPCCT in preclinical investigation can be employed with the same protocol setting as in clinical practice which means all preclinical studies can be translated for human imaging. Molecular imaging with SPCCT has paved the way for several imaging applications, using contrast agents and nanoparticle-based compounds.

The present chapter will discuss the potential role of SPCCT in molecular imaging for cancer, infectious disease, cardiovascular disease and orthopaedic applications and provide a comparison to the existing modalities.

3 Molecular Imaging Applications

3.1 Infectious Disease: Pulmonary Tuberculosis

In 2019, 10 million new cases of the highly infectious lung disease, pulmonary tuberculosis (TB), were recorded by the World Health Organization, and 1.2 million TB-related deaths occurred [26]. Drug-resistant strains accounted for 3.3% of deaths [27, 28]. Efforts to curb TB and TB drug resistance remain a public health crisis. Novel therapies have resulted in a decline in TB incidence of approximately 2% per year. However, incidence rates must decrease quicker to reach the goal set by the United Nations' Sustainable Development Goals (SDG) [26].

Drug efficacy studies of novel therapies are plagued by poor patient compliance due to lengthy and complex treatment protocols and long periods of follow-up. Consequently, predictive preclinical imaging plays a critical role in facilitating the development of novel drug therapies. Preclinical efficacy studies using animal models of TB disease are key to supporting the rationale of clinical trials involving patients [29]. Micro-CT is the most commonly used imaging technique in preclinical studies to assess disease burden and drug efficacy. However, studies using micro-CT cannot be directly translated to clinical human imaging. Thus, the use of an imaging modality such as SPCCT, which is capable of both preclinical and clinical applications, would result in a more rapid translation.

A proof-of-concept study by C. Lowe et al. [129] presented SPCCT as a clinically translatable, molecular imaging tool by assessing the uptake of iodine-based contrast into an ex vivo murine model of TB. Shown in Fig. 1, the excised mouse lungs were placed in paraffin and imaged using a standard micro-CT (SuperArgus) and the contrast-enhanced TB lesions quantified. The same lungs were imaged using a small-bore SPCCT research scanner (MARS Microlab 5 × 120, MARS Bioimaging Limited, New Zealand), as shown in Fig. 2. Iodine and soft tissues (water and lipid) were materially separated, and iodine uptake quantified.

Quantification was performed via a well-established thresholding method which separated background, healthy and diseased tissue. Histology was used to provide ground truth and check for correlation between images and histological cuts. Lesions located by micro-CT and SPCCT were also visualized on histology. Histology showed no lesions were found in the healthy lungs. The volume of the TB infection quantified by spectral CT and micro-CT was found to be 2.96 and 2.83 mm^3, respectively. This proof-of-concept study showed that SPCCT could be used as a predictive preclinical imaging tool for the purpose of facilitating drug discovery and development. Also, as this imaging modality is available for human trials, all

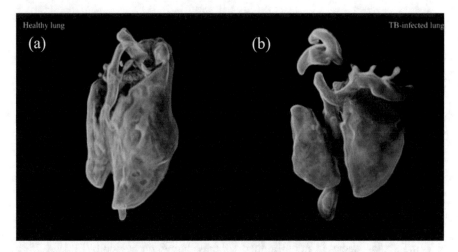

Fig. 1 Micro-CT 3D visualization of (**a**) healthy lung, (**b**) TB-infected lung

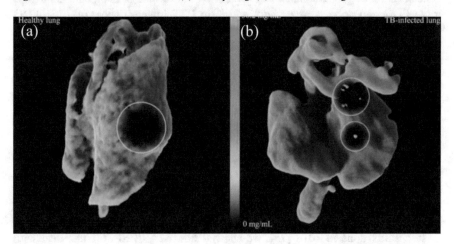

Fig. 2 SPCCT 3D visualization of (**a**) healthy lung in which the white circle indicates no iodine uptake as there are no lesions in the healthy lung and (**b**) TB-infected lung in which the white circle indicates iodine uptake by TB infection lesions

applications are translatable to human imaging. In this study, we used iodine as the common contrast agent in clinical CT imaging to locate the TB infectious lesions. To differentiate iodine from soft tissue, the energy thresholds have been set at 20, 27, 34 and 45 keV in charge summing mode (CSM) at a spatial resolution of 90 micron. In conclusion, SPCCT enables a deeper understanding of infectious lung diseases using targeted pharmaceuticals and intrinsic markers. This would improve the efficacy of therapies by measuring drug delivery and response to treatment in animal models, and later in humans.

3.2 Cancer

Worldwide, cancer is one of the leading causes of morbidity and mortality. Late diagnosis is a major contributing factor to treatment failure [30–32]. Cancer treatment may differ depending on the type of cancer and the cancer cell line. A further challenge in treating cancer is ensuring adequate drug delivery [33]. Most cancer treatments are selected based on randomized clinical trials and may not be fully personalized to the patient and their tumour biology. At present, biopsies are used to determine cancer lines and tumour heterogeneity. Nonetheless, biopsy is an invasive method and is limited by the number and size of samples.

Molecular imaging is an emerging field that has the potential to enable personalized cancer care. PET is the most advanced molecular imaging technology, which is used routinely to diagnose and grade cancers. Currently, 18F-fluoro-2-deoxy-D-glucose (18F-FDG) is widely used for PET cancer imaging; however, this molecular tracer is not designed to only target cancer cells [34]. A high concentration of glucose transporter protein on the surface of malignant cells traps more FDG and appears as a "hot" region in PET images. Therefore, 18F-FDG is taken up by tumours regardless of type which means the PET scan does not provide precise information to select the best treatment. In addition, 18F-FDG is taken up by some disorders, such as inflammation and hyperplastic bone marrow. Furthermore, some tumours are not PET-avid and cannot be reliably visualized by PET. In contrast, SPCCT using functionalized nanoparticles (NPs) has the potential to detect the specific cancer cells and pave the way for personalizing the treatment and monitoring tumour heterogeneity at all relevant sites [25, 35]. As an example, coincident breast cancer and lymphoma are rare clinical conditions, but the rate of misdiagnosis is high. Thus, imaging modalities that can distinguish these two cancers can facilitate treatment significantly. The studies proved that the uptake of 18F-FDG rises significantly for both breast cancer [36] and lymphoma [37]; therefore, in a single PET/CT scan, there is no distinction between these two cell lines [38].

In a study done in 2018 [25], we showed how MARS SPCCT can deal with this challenge and distinguish between two cancer cell lines. The crossover study was carried out using MARS SPCCT in conjunction with functionalized gold NPs (AuNPs) to target specific cancer cell lines. AuNPs are considered an excellent candidate for SPCCT imaging as they are biocompatible, exhibit lower toxicity in vivo, can circulate longer, have more flexibility with respect to functionalization and targeting, attenuate X-rays strongly and also have a high K-edge value while still being within the diagnostic energy range [39–42]. HER2-positive human breast cancer cell line (SK-BR3) and CD20-positive human B-cell line (Raji) were selected and incubated with gold-labelled monoclonal antibody, Herceptin (to target the HER2-positive cancer cell line) and rituximab (to target CD20-positive cancer cell line). Results showed that the binding of AuNPs to SK-BR3 cells was eight times higher when AuNPs were functionalized with Herceptin, as compared to AuNPs functionalized with rituximab (Fig. 3). In contrast, Raji cells took up

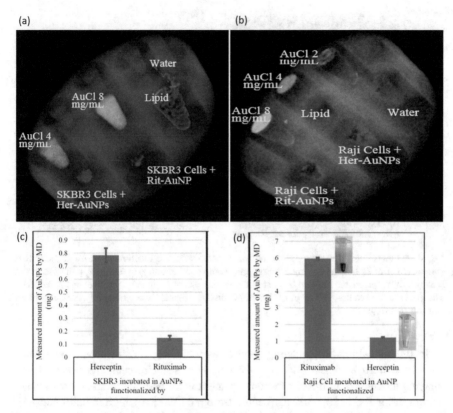

Fig. 3 (**a**, **b**) MARS MD images of phantoms that indicate Au (yellow), lipid (pink) and water-like material (grey). Depth of colour indicates material concentration. (**c**, **d**) Measured amount of targeted and nontargeted AuNPs uptaken by HER2-positive cells (Herceptin-gold targeted to HER2+ cell) and Raji cells (rituximab-gold targeted to Raji cell)

rituximab-gold-labelled nanoparticles six times more than Herceptin-gold-labelled nanoparticles. The finding of this study provided proof of principle that MARS SPCCT can detect and measure specific gold-labelled monoclonal antibodies to enable accurate diagnosis of tumour type.

As well as diagnosing the type of cancer, it is important to detect it in its early stage, as early detection can make the treatment more effective. Angiogenesis, the growth of new blood vessels, is crucial to tumour growth and cancer development. These new vessels are known as "leaky vessels" because of the broader gaps in the endothelium compared to normal vessels and form at the early stage of cancer and tumour progression [43, 44]. Angiogenesis is the prognostic factor in several cancers including breast and lung cancer [45–48]. In addition, evaluation of these disorganized blood vessels can provide information regarding the efficacy of treatment response.

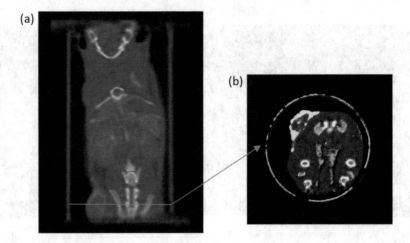

Fig. 4 A Lewis lung tumour mouse model was injected with non-functionalized AuNPs and then euthanized after 24 hours. (**a**) Single energy grey-scale coronal image of a mouse with gold in the kidneys, liver, spleen and tumour. Gold is intravascular except for the tumour, liver and spleen. (**b**) The gold (yellow) in the vicinity of the tumour is distinguishable from the bone (white), fat (blue) and soft tissue (grey)

In preclinical studies, imaging modalities including CT, ultrasound and MRI have been used to detect and evaluate angiogenesis. However, the use of these imaging techniques in the clinical environment is limited due to spatial resolution in conventional CT (500–800 μm) and ultrasound (500 μm), high radiation dose in CT perfusion and low sensitivity in MRI [49]. Although dynamic contrast-enhanced (DCE) MRI shows promising results for measuring response to treatment for several types of cancer, physiological motion of the organs, patients with metallic implants and a lack of validation methodology all pose significant obstacles to using DCE-MRI in clinical practice [50].

The stated limitations of current clinical imaging can be overcome using SPCCT which has a high spatial resolution, delivers low radiation dose [19] and has no limitations to image metallic implants [51]. A pilot study was designed to assess the ability of MARS SPCCT to detect angiogenesis [52, 53]. MARS SPCCT was used to scan a mouse tumour model of Lewis lung carcinoma (LL/2) that was injected with 15 nm AuNPs. Through accurate differentiation between AuNPs as a contrast agent, soft tissue, lipid and bone (Fig. 4), SPCCT can assess angiogenesis and the tumour separately. SPCCT also enables quantification of AuNPs within angiogenesis which is an important feature to use NPs for drug delivery or measuring the treatment response. The result indicated that SPCCT technology allowed for better characterization of tumour angiogenesis, which could lead to faster development of targeted cancer therapies, monitoring drug delivery and disease response, as well as treatment adjustment.

Also, we examined how the concentration and size of AuNPs influence cellular uptake. SKOV3 and OVCAR 5 ovarian cancer cells were incubated with four sizes

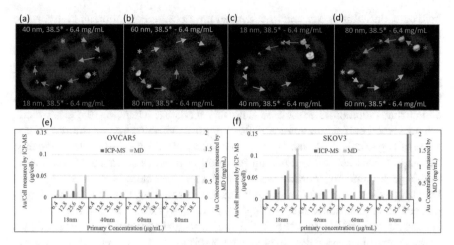

Fig. 5 MARS MD images of OVCAR5 cells (**a, b**) and SKOV3 cells (**c, d**) with AuNPs. (**e, f**) indicate gold/cell (calculated based on ICP-MS and DNA quantification) along with quantification of gold by MARS spectral imaging. Arrows indicate lower to higher concentration

of AuNPs (18, 40, 60 and 80 nm) at four concentrations (6.4, 12.8, 25.6 and 38.5 μg Au/ml). Results showed that uptake of AuNPs varied with cell type and size of AuNPs. It was found that 18 and 80 nm were the optimal sizes for detecting SKOV3, but AuNPs of the same size and concentration showed very low uptake by OVCAR5 (Fig. 5). The results of this study have been used as the basis for selecting an optimal nanoparticle to use as a targeted contrast agent for spectral CT imaging [54].

3.3 Characterization of Atherosclerotic Plaque

For decades, cardiovascular diseases (CVDs) have been the main cause of death and disability, accounting for one in three deaths, attributed to coronary artery diseases (CADs) and strokes [55]. The most common cause of CVD is atherosclerosis, which is an underlying progressive and inflammatory process occurring in the arteries leading to plaque formation. In advanced stages of atherosclerosis, the plaques rupture which form blood clots that obstruct arteries further downstream. As a consequence, patients experience strokes and heart attacks due to the lack of blood supply to the brain and heart [56].

A key feature of atherosclerosis is thickening of the arterial walls, which is caused by the deposition of several materials such as lipids, cellular debris, calcium and other fibrous materials. Current imaging modalities used to visualize atherosclerotic plaques in clinical practice include computed tomography angiography (CTA) and magnetic resonance angiography (MRA). Although both modalities have their strengths, they both lack the spatial resolution and molecular discernment to detect the features of vulnerable atherosclerotic plaques, such as a large necrotic lipid core, thin fibrous cap, micro-calcifications, neovascularization and intraplaque haemorrhage [57–60]. CTA can detect small calcifications, but large calcifications make it difficult to estimate the degree of stenosis due to blooming of the calcium signal. This makes it too difficult to distinguish the blocked region from the imaging artefact. CTA also has poor soft tissue contrast, due to its reduced spatial resolution. Injected contrast agents are needed to visualize blockages, and a high dosage of ionizing radiation is associated with this modality. MRA, on the other hand, provides good soft tissue contrast, allowing direct visualization of stenosis, but is not sensitive to calcifications. Injected contrast agents are also used in this modality, although it does not utilize ionizing radiation to produce an image. A significant challenge for imaging aortic diseases such as atherosclerosis with MRA is the long acquisition time that can cause artefacts due to respiratory motion, leading to reduced contrast-to-noise ratio and spatial resolution, and the need for ECG-gating to correct artefacts [61]. In addition, the morphological approach of these techniques does not provide information on plaque metabolism, inflammation and progression, which are markers of plaques' vulnerability.

PET and SPECT are capable of providing functional information on cellular changes relating to the risk of plaque rupture and evaluating the determinants of plaque vulnerability through specific cellular and biochemical changes [62]. PET imaging uses the 18F-FDG radiopharmaceutical which is taken up in metabolically active regions through glucose metabolism and accumulates in atherosclerotic plaques. The limitations of PET imaging include poor spatial resolution, radionu-clide radiation exposure and the short half-life (110 minutes) of 18F-FDG. This presents a logistical challenge for radioisotope supply to clinics located far from the place of radioisotope production. SPECT imaging of atherosclerotic plaque uses technetium-99 m-labelled (99mTc) radiotracers to investigate myocardial perfusion and viability in coronary plaques. The half-life of 99mTc is 6 hours, making SPECT more widely accessible and less expensive than PET. However, this modality suffers significantly with poor resolution and prevalent attenuation artefacts.

Combining PET/MRI can further enhance visualization of soft tissue boundaries in multimodality imaging [63]. Although multimodality hybrid imaging can con-front the issues of poor resolution associated with PET and SPECT, this technique is extremely costly.

We have performed preclinical studies focused on cardiovascular imaging appli-cations using SPCCT to demonstrate how SPCCT could pave the way for better diagnosis of cardiovascular diseases. Phantom studies involving concentrations of iron (ferric nitrate) and calcium (hydroxyapatite) have been designed to distinguish two spectrally similar materials by analysing their attenuation coefficients across the energy thresholds [24]. Iron and calcium are considered inflammatory compo-nents of unstable plaque through intraplaque haemorrhage and micro-calcification, respectively, and are important for identification of vulnerable plaques that are prone to rupture [64]. This work led to scanning an excised carotid plaque from a stroke patient, which identified a collection of iron signals surrounded by regions of calcifications, lipid and soft tissue (Fig. 6). The site of rupture was also identified by ulceration in the wall of the plaque.

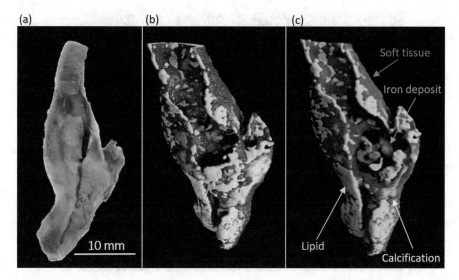

Fig. 6 A surgically removed carotid plaque taken from a 74-year-old non-diabetic male smoker treated following a stroke. (**a**) Excised specimen, as photographed under white light using a macro-lens (**b**) MARS image of the plaque's surface (**c**) MARS material image. Calcium-rich regions are shown in white, lipid-rich are in yellow/white and water-dominated tissue in red. Subtraction of the water, lipid and calcium channels allows the iron-rich region to be visualized showing the extent of the intraplaque haemorrhage

ApoE$-$/$-$ mice are an adaptable atherosclerosis model due to their delayed lipoprotein clearance, leading to spontaneous lesion development. Aged ApoE$-$/$-$ mice were fed a chow diet and imaged using MARS SPCCT. The imaging identified regions of calcifications within the ascending aorta and aortic root, which was confirmed by histological analysis of the aortas (Fig. 7).

A separate experiment, involving an excised carotid plaque sutured into the popliteal fossa of a lamb knee sample, was performed to replicate the conditions for a peripheral artery disease patient scan. The scan parameters were standardized to achieve optimal soft tissue contrast while reducing the effects of beam hardening. The unique regions of carotid plaque calcification and the soft tissue components of the carotid plaque were able to be distinguished within the lamb knee tissue (Fig. 8). The experiment was followed by a clinical trial scan of the peripheral artery disease in the patient, which revealed calcifications in the posterior tibial and peroneal artery. The fibula and tibia bone were visualized with no streaking artefacts, and the inner wall of the artery was clearly identified as soft tissue (Fig. 9).

In addition, we developed our application to detect atherosclerosis at early stages using the SPCCT in conjunction with functionalized nanoparticles. As aforementioned, molecular imaging with SPCCT has paved the way for several imaging applications, using contrast agents and nanoparticle-based compounds. Cell tracking is one such application whereby cells can be tagged with contrast

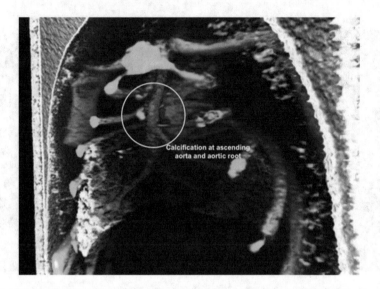

Fig. 7 MARS volumetric visualization of an ApoE−/− mouse with blue representing bone and calcium, and yellow representing lipid. The circle indicates the regions of calcification along the ascending aorta and aortic arch

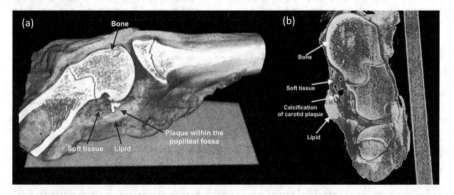

Fig. 8 (**a**) MARS 3D image and (**b**) axial view of an excised carotid plaque sutured into the popliteal fossa of a lamb knee where white represents bones and calcification, red represents soft tissue and yellow represents lipid

agents or nanoparticles, and their build-up in regions of interest can be tracked via imaging. Contrast agents and nanoparticles have several physical and chemical properties that can be used to our advantage. Nanoparticles can be potentially used in atherosclerosis management as they can be targeted specifically to plaques [65].

In this case, monocytes were labelled with 11-mercaptoundecanoic acid-coated gold nanoparticles (11-MUDA AuNPs) and injected into ApoE−/− mice, followed by scanning using a SPCCT. The images show the accumulation of the monocytes at the site of interest, i.e. the plaque in the aortic arch, suggesting that monocytes

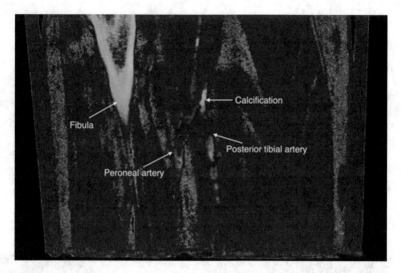

Fig. 9 Coronal view of the fibula and tibia bone, white representing bone and calcification, yellow representing lipid and red representing soft tissue

are recruited into atherosclerotic regions [66]. These images also show the potential of the MARS imaging system to differentiate between various components, such as soft tissue, hydroxyapatite (HA) and gold (Fig. 10).

3.4 Bone Health

The human skeleton is composed of bone, cartilage and ligament tissue. Skeletogenesis is the process of bone formation during the prenatal stage; however, the same process can occur in the postnatal stage as bone regeneration [67]. Unlike the soft tissue healing process, which occurs by generating scar tissue, the bone heals through the formation of new bone that does not generate fibrous tissue [67, 68]. Bone healing is followed by various and complex interactions of cells and connective tissues [69–72]. Bone consists of two major tissues, namely, cortical and trabecular bone. The mineral content within these tissues is an important indicator of bone quality [73]. Additionally, the structure of these tissues gives insight into the bone strength and is reported to be an informative factor to detect bone disorders and fractures. Bone mineral properties are key factors to study bone-related diseases and can be used in the field of orthopaedics [74]. The bone mineral density (BMD) in common fracture sites, such as the femoral neck, spine and wrist, is assessed using dual-energy x-ray absorptiometry (DXA) [75]. However, this modality only provides the areal BMD without considering the depth factor, where there are several tissues which overlap [76]. Additionally, this imaging technique is not able to resolve bone tissues (cortical and trabecular bone), so if the mineral

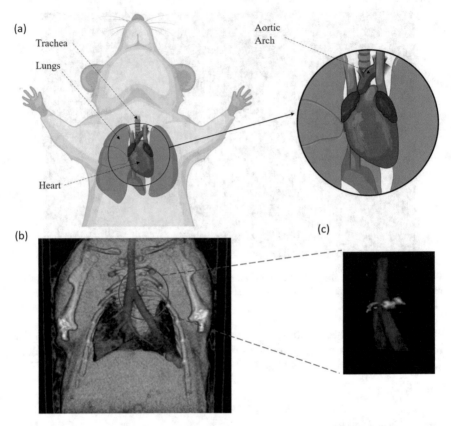

Fig. 10 (**a**) Heart and aortic arch anatomy. (**b**) MARS fused MD images including energy image at energy image (pink), bone (white) and soft tissue (cyan). (**c**) MARS fused MD images of aortic arch including energy image at energy image (pink), bone (white), and AuNPs (yellow)

loss happens only in trabecular tissue, it cannot be detected [76, 77]. Although BMD measurement is considered as an important criterion for bone strength, and fracture risk assessment, this criterion only accounts for partially defining the bone strength [78]. Indicating the bone structure should also be taken into account while assessing bone strength [79, 80]. Quantitative computed tomography (QCT) provides 3D images of a bone with the ability to discriminate between cortical and trabecular bone [81]. However, the resolution is not high enough to resolve detailed micro-structure [81]. Peripheral quantitative CT (pQCT) provides BMD measurements and quantitative assessment of the cortical bone in the distal tibia and radius [82]. Although pQCT has a lower radiation dose compared to QCT, it has a relatively slow scanning process, increasing the risk of motion artefact, and cannot be used in common osteoporotic sites such as the proximal femur and spine [83]. High-resolution peripheral quantitative CT (HRpQCT) can resolve the cortical and trabecular bone [84] with an effective radiation dose that is comparable with multi-detector CT. However, partial volume averaging will occur in the bone with 50 μm thickness [76]. In addition, this imaging modality cannot be used in the common osteoporotic sites and needs a dedicated CT scanner which makes it expensive [76]. Micro-CT provides detailed structural information along with the BMD measurements, but this modality is restricted to preclinical research [85]. MRI can be used as an imaging technique with no radiation dose to provide information about the bone marrow [76]; however, the trabecular thickness measured from MRI was found to be three times greater than the measurements from HRpQCT [86]. Quantitative ultrasound can measure the density and structure of bone using broadband ultrasound attenuation (BUA) and ultrasound velocity (UV), respectively [87–89]. However, this technique is limited to superficial bones due to the lack of equipment required for evaluating the density and structure of bone underneath the thick layers of muscle [90].

Therefore, an imaging modality that provides high resolution and accurate quantitative measurements of bone tissues, as well as the ability to image all osteoporotic sites, should improve detection of bone disorders. MARS spectral photon-counting CT is able to assess the volumetric bone mineral density and bone architecture at 70–150 micron voxel size simultaneously [91]. This technology provides site-specific BMD regardless of the position of the bone. The voxel size is reasonable to resolve the trabecular bone, although it is not small enough to visualize the fine structures that can be observed with micro-CT. However, unlike micro-CT, this imaging technology is not destructive and imposes less radiation dose compared with conventional CT. Bone mineral content and morphology obtained from SPCCT showed comparable results to DXA and HR-pQCT, respectively, in a preclinical research study [92].

Figure 11 shows the quantification of bone mineral content using MARS spectral photon-counting CT and its comparison with the reference method iDXA. These measurements were acquired every 2 weeks after surgically induced damage in sheep tibia. Figure 12 shows the 3D calcium map of healing within the damaged area using MARS spectral photon-counting CT.

Qualitative assessment of monitoring the damaged area healing allows the clinicians to assess the type of fracture healing process. MARS SPCCT images were compared with the clinically available x-ray technologies. Images demonstrated the capability of the radiograph, clinical CT and SPCCT to resolve trabecular bone from cortical bone. This criterion offers the opportunity to evaluate the process of healing in cortical and trabecular bone separately and simultaneously. The granulation tissue formation, an early sign of the healing process, was traceable using clinical CT and SPCCT images. However, due to the smaller voxel size in SPCCT, the lines of the healing region can be better defined compared to CT images. DXA images cannot resolve the trabecular and cortical bone; therefore, the results from this imaging modality can only represent the overall bone mineral content. In addition, the DXA images lack adequate spatial resolution and do not provide information about the periosteal reaction or callus formation. Figure 13 shows the tissue inside the hole in the bone has a radiographic appearance similar to soft tissue. The formation of the granulation tissue between the induced hole and beyond the periosteal sites

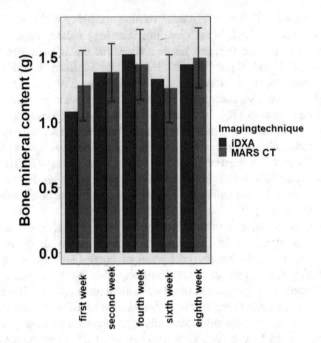

Fig. 11 Quantification of bone mineral content in the area of surgically induced damage in sheep tibia every 2 weeks

can be traced in clinical CT and spectral photon-counting CT images. This feature is evidence of bone healing. Visual assessment of healing is not possible in the DXA images. Figure 14 shows the observation of the lamellar bone in clinical CT and SPCCT images after 8 weeks. All images obtained 8 weeks after surgery demonstrated periosteal bone formation.

Furthermore, we showed that the SPCCT can diagnose microfracture using targeted hafnia nanoparticles (HfO_2) [93]. Bone microdamage related to tensile and shear force is a common physiological phenomenon and is important for bone remodelling and maintaining bone integrity [94]. Particularly with certain pathological conditions which lower bone mineral density such as osteoporosis, bone microcracks may extend to develop into stress and fragility fractures [95, 96]. Earlier diagnosis of this microdamage along with bone mineral content should help to prevent a significant number of fractures [97], thereby reducing healthcare cost and severity. Currently, microdamage is only diagnosed with destructive techniques such as histomorphometry [98]. Detection of this kind of pathology with conventional CT scan is limited due to partial volume effects related to poor spatial resolution and the inability to differentiate between bones with contrast agents in energy integrating CT systems. MRI has also limited resolution and cannot demonstrate bony detail. Therefore, none of the current technology is sufficient to image these minute defects in bone.

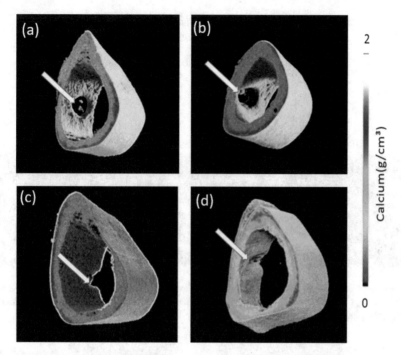

Fig. 12 Volumetric 3D calcium maps (**a**) after 2 weeks of treatment, (**b**) after 4 weeks of treatment, (**c**) after 6 weeks of treatment, and (**d**) after 8 weeks of treatment. The arrows show the area of healing bone where the calcium density is increasing

The MARS imaging system can differentiate between calcium and other high-Z contrast media with high resolution, which can be employed for detection of molecular changes in bone. Microdamage causes an increase in calcium ions in the matrix; therefore, calcium-specific nanoparticle contrast agents can accumulate in this region. Contrast uptake in the region of microdamage with hafnium ($Z = 72$) nanoparticles has been demonstrated using MARS technology (Fig. 15) [93]. Hafnium has a favourable K-edge for MARS spectral CT, and HfO_2 nanoparticles are undergoing clinical trials [93]. The potential of MARS to demonstrate other nanoparticle contrast agents has also been demonstrated [25]. The imaging of calcium ion concentrations in specific bone sites using specific ligands has applications for other spectral molecular applications such as bone healing and physiological conditions of the bone. The simultaneous quantification of bone mineral content, together with the assessment of the site and location of microdamage, could provide reliable data for bone health assessment and fracture risk prediction.

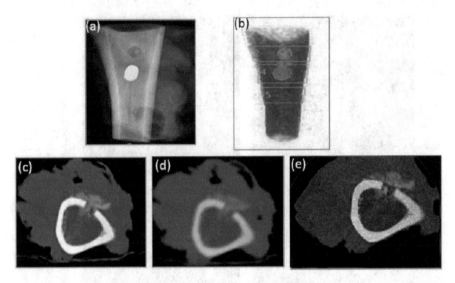

Fig. 13 Images of the sheep sample obtained after 2 weeks of induced hole. (**a**) Plain radiograph. (**b**) iDXA. (**c**) Conventional CT. (**d**) Conventional dual-energy imaging using fast kVp switching. (**e**) Spectral photon-counting CT

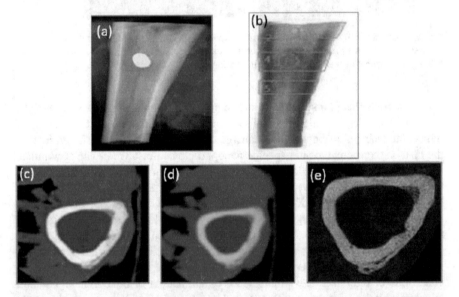

Fig. 14 Images of the sheep sample obtained after 8 weeks of induced damage. (**a**) Plain radiograph. (**b**) iDXA. (**c**) Conventional CT. (**d**) Conventional dual-energy imaging using fast kVp switching. (**e**) Spectral photon-counting CT

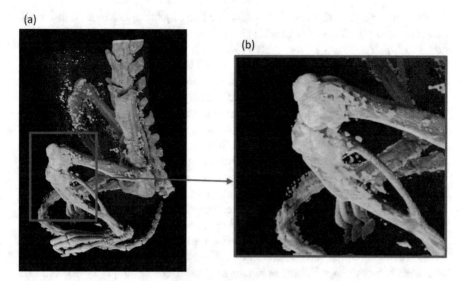

Fig. 15 (**a**) MARS 3D volumetric images of the rat injected with HfO$_2$, red box (**b**) indicates the site of induced microfracture and the accumulation of targeted HfO$_2$. White represents bone (calcium channel), cyan to orange represent Hf (low to high concentration) and red represents soft tissue

3.5 Cartilage Health

Another important molecular application of MARS spectral CT is the early and accurate assessment of articular cartilage health. Using this quantitative molecular technique, articular cartilage degradation can be diagnosed earlier than morphological changes occurring in joints, thereby contributing to a change in patient management towards early interventions.

Osteoarthritis (OA) is a painful degenerative joint disease characterized by the degeneration of the articular cartilage, which in turn leads to loss of mobility and decreases quality of life. Earlier diagnosis of the biochemical status of abnormal cartilage can be performed by the assessment of changes in the extracellular matrix of the cartilage [99].

The quantification of contrast uptake by negatively charged sulphated glycosaminoglycans (sGAG), an important component in the extracellular matrix of cartilage, helps to diagnose abnormality in the earlier stages before morphological changes can happen [100, 101]. GAG is the marker of cartilage degradation in the early phase of osteoarthritis [102]. The amount of GAG depends on cartilage thickness with the superficial layer containing less GAG relative to the deep layer [103, 104]. In osteoarthritis, the GAG level in the superficial zone is further deteriorated [105]. This zonal distribution of GAG is traditionally measured by destructive techniques such as histology [106] or biochemical methods [107].

Plain radiographs are a common diagnostic imaging technique used to indirectly detect the complications of OA such as joint space narrowing, osteophyte formation and subchondral sclerosis and cysts [108, 109]. These radiographic findings are late-stage changes, and by the time they are visible, permanent joint damage has occurred. Magnetic resonance imaging can produce excellent soft tissue detail to demonstrate the morphological lesions in cartilage and subchondral bone [110]. Delayed gadolinium-enhanced magnetic resonance imaging of cartilage (dGE-MERIC) is an MRI-based post-contrast technique for the quantitative assessment of cartilage health. In this technique, after the administration of anionic gadolinium contrast agents through intravascular or intraarticular route, the zonal distribution of GAG can be assessed. There is a higher level of contrast uptake in regions with low GAG (superficial zone) and lower uptake in regions with high GAG (deeper zone) [111]. T1ρ is another MRI-based technique which can indirectly assess the GAG content by using a special pulse sequence. In this technique, higher T1ρ value indicates low GAG content [112]. Similarly, T2 mapping is another MRI-based technique which can demonstrate cartilage degradation based on assessment of orientation of collagen fibres as a marker of cartilage health [113]. Sodium MR imaging of cartilage using 7 T MRI is also under investigation in the research stages [114]. However, assessment methods using MRI are still limited due to lower spatial resolution. High-resolution equilibrium partitioned imaging of cartilage (EPIC) with micro-CT using ionic contrast agents has shown contrast agent uptake that is inversely proportional to GAG content [115, 116]. However, it is difficult to separate the contrast uptake in cartilage with subchondral bone with this CT scan technique in the absence of spectral information.

MARS spectral CT scan can quantify the GAG distribution in a similar pattern to dGEMERIC, with high-resolution images and material-specific images to differentiate cartilage from subchondral bone. Thus, using this technology, the cartilage layer can be properly segmented and biochemical changes in thinner articular cartilages can be assessed. MARS imaging can be performed with cationic or anionic contrast agents; however, various studies have shown better quantification potential with cartilage-specific cationic contrast agents [100, 117]. The ability of the MARS imaging system to quantify both iodine- and gadolinium-based anionic contrast has been already demonstrated [118]. Various studies have validated that the contrast concentration measured from MARS is inversely proportional to GAG distribution in the cartilage [118, 119]. The relationship between MARS material maps and histology is demonstrated in Fig. 16.

The ability of the MARS imaging system to image multiple contrast agents simultaneously has been already demonstrated [120]. Therefore, imaging can also be integrated with a combination of non-ionic contrast from one material and ionic contrast from another material to increase diagnostic certainty in the case of early OA patients. The distribution of gadolinium contrast (Dotarem) in proportion to GAG distribution in healthy bovine knee cartilage is demonstrated in Fig. 17. The MARS system has now been employed for the clinical trial; therefore, translational research to determine the clinical pathway of cartilage imaging is ongoing.

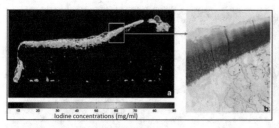
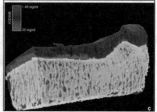

Fig. 16 Biochemical measurement of cartilage health in human OA knee sample using ionic iodinated contrast agents. The iodine contrast concentration in proportion to glycosaminoglycan distribution is demonstrated in (**a**) in the iodine map image. The histology of GAG distribution is demonstrated with (**b**), which closely matches the region marked with box in (**a**). The iodine contrast distribution in different layers of cartilage along with subchondral bone is demonstrated in 3D image (**c**). (Image courtesy from Ref. [119])

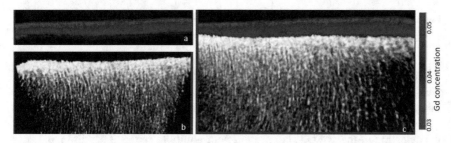

Fig. 17 Biochemical measurement of cartilage health in healthy bovine patella sample. (**a**) Gadolinium map showing the distribution of contrast along the different layers of cartilage thickness. (**b**) Hydroxyapatite map demonstrating subchondral bone and trabecular bone. (**c**) Both maps were fused; however, the cartilage layer is clearly distinguished from the subchondral bone layer

3.6 Crystal-Induced Arthropathies

Crystal arthritis (CA) is an articular disorder caused by the formation of different types of crystals in the joints. The main symptoms of the different CA are similar, mainly characterized by an acute inflammation in the joint. The most common forms of CA are gout, caused by the formation of monosodium urate crystals (MSU), pseudogout due to the formation of calcium pyrophosphate (CPP) and basic calcium phosphate arthritis, mainly due to hydroxyapatite (HA) crystal deposition [121].

The identification of the type of crystal present in the joint is clinically significant as patient management depends on it. Today, the gold standard diagnostic test is performed via aspiration of the synovial fluid or tophus and analysed using polarized light microscopy, x-ray diffraction or Raman spectroscopy [23, 122, 123]. The aspiration of the synovial fluid implies a risk of complications, such as introducing infection into the joint. Besides, spectroscopy techniques are not always accessible due to availability of the technology and the expertise, and polarized light

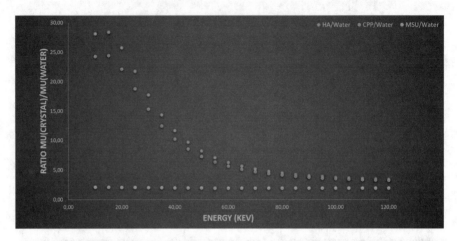

Fig. 18 Attenuation coefficient of crystals relative to water as a function of the energy. (Data collected from the National Institute of Standards and Technology (NIST) XCOM database [128])

microscopy is operator-dependent. Therefore, a non-invasive, accessible method for the diagnosis of CA would make a significant difference for patients suffering from this pathology.

Traditional methods such as plain radiography, CT and US can be helpful for the detection of the crystals, but present poor ability to separate the different types of crystals [123–125]. Although US presents a good CPP detection in some joint structures, such as hyaline cartilage or fibrocartilage, further tests such as synovial fluid analysis are required for a definitive diagnosis [122, 126].

In recent years, new advanced imaging techniques such as dual-energy computed tomography (DECT) or spectral photon-counting computed tomography (SPCCT) have shown good potential for the diagnosis of CA [23, 122, 124, 127]. DECT and SPCCT measure the attenuation profile at different energies to identify the different crystals. CPP, HA and MSU crystals present different x-ray photon attenuation profiles across the energy range used in medical imaging as shown in Fig. 18.

The profile for MSU differs from CPP and HA which therefore provides an ideal opportunity to use DECT or SPCCT to identify CA as a result of MSU crystal deposits. Figure 19 shows the result obtained from scanning an excised finger with a gouty subcutaneous tophi using plain radiography, DECT and SPCCT (MARS imaging system) [23]. Plain radiography showed bone erosion and changes in the soft tissue characteristic of MSU deposition. Both DECT and SPCCT presented the same anatomical characteristics as plain radiography, with the inclusion of identification of MSU crystals in the tissues (colour coded in green in the images). However, SPCCT presented finer detail of the crystal distribution and higher MSU volume. The presence of MSU was confirmed by polarized light microscopy.

For CPP and HA, the diagnosis is more challenging because the attenuation profiles of CPP and HA are close across the diagnostic energy range (Fig. 18). However, studies have shown the potential for identifying these crystals [23, 127].

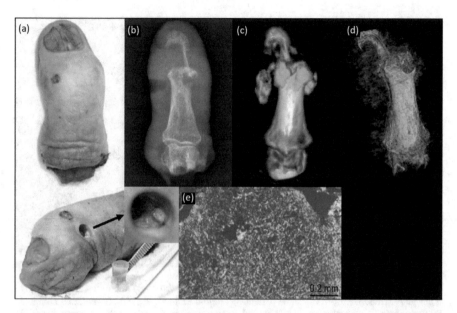

Fig. 19 (**a**) An excised gouty finger with a core (arrow) revealing subcutaneous tophus. (**b**) Plain x-ray showing bone erosions with overhanging edges and soft tissue changes consistent with tophus. (**c, d**) Dual-energy computed tomography image (**c**) and multi-energy spectral photon-counting computed tomography (SPCCT) image (**d**), both depicting monosodium urate (MSU) crystal deposits, with SPCCT detecting finer detail and higher MSU volume. (**e**) Polarized light microscopy of a sample obtained from the core, confirming the presence of negatively birefringent MSU crystals. (Image retrieved from Ref. [23])

Figure 20 presents the results obtained from an excised femoral condyle of a patient suffering from CA. Results showed that different imaging systems could be used to detect the crystals within the cartilage (white arrow in Fig. 20), but there was a clear difference in imaging quality between the modalities. DECT and ultrashort echo time magnetic resonance imaging (UTE MRI) could quantify calcium crystal deposits but missed the crystals at the surface of the cartilage (yellow arrow in Fig. 20). SPCCT could quantify the calcium crystal deposition, and it was the only imaging modality that detected HA crystal deposition along with CPP [23]. The presence of both CPP and HA in the sample was confirmed with Raman spectroscopy.

Those preliminary studies indicated SPCCT can provide increased specificity in the crystal identification and higher sensitivity than DECT, with high spatial resolution. Therefore, this imaging modality presents good potential to improve CA diagnosis non-invasively. Furthermore, the ability to quantify calcium deposition could help clinicians to monitor the evolution of the pathology. However, further studies, including clinical trials, need to be conducted to assess the place of SPCCT imaging in the CA diagnostic process.

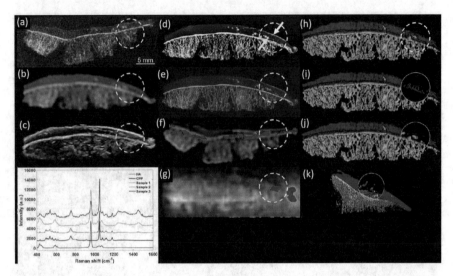

Fig. 20 Multimodality comparison of ex vivo images of the (**a–j**) femoral condyle and (**k**) tibial plateau samples from the same patient: (**a**) digital radiography, (**b**) conventional computed tomography (CT), (**c**) ultrasound, (**d**) digital mammography, (**e**) micro-CT, (**f**) dual-energy CT (DECT), CPP is colour coded in red (**g**) ultrashort echo time (UTE) magnetic resonance imaging (MRI), CPP is colour coded in red (**h–k**) Spectral photon-counting CT (SPCCT), CPP colour coded in red, HA in white, water in blue. (Bottom left) Raman spectroscopy was used as the reference standard (synthetic CPP and HA spectra are colour coded in red and blue, respectively) and confirmed the coexistence of CPP and HA crystals in two of the three cartilage scrapings/biopsies from the tibial plateau sample. (Image retrieved from Ref. [127])

4 Summary

Molecular imaging in medicine is used to visualize, characterize and quantify pathophysiological processes in vivo. The main advantage of molecular imaging in vivo is its ability to identify disease processes without invasive surgical procedures and biopsies. Such imaging allows for more tailored treatment. This chapter reviews the current molecular imaging modalities used in the clinical setting and preclinical studies, as well as their limitations to study disease and drug delivery. CT and MRI suffer from low contrast sensitivity and moderate contrast sensitivity, respectively. Depth of ultrasound penetration makes US applicable to the surface organs only. PET and SPECT are associated with added radiation risk and technical disadvantages, and micro-CT is limited in its ability to translate study findings to humans. SPCCT offers an alternative method for molecular imaging using intrinsic biomarkers and non-radioactive pharmaceutical agents. Thus, SPCCT may open the door to a broader range of medical investigations and increase patient access to advanced diagnostics.

Acknowledgements This project was funded by the Ministry of Business, Innovation and Employment (MBIE), New Zealand, under contract number UOCX1404, by MARS Bioimaging Ltd and the Ministry of Education through the MedTech CoRE. The authors would like to acknowledge the Medipix2, Medipix3 and Medipix4 collaborations. Also, they would like to take this opportunity to acknowledge the generous support of the MARS Collaboration.

MARS Collaboration Sikiru A. Adebileje[a,e]; Steven D. Alexander[a]; Maya R. Amma[a]; Marzieh Anjomrouz[a]; Fatemeh Asghariomabad[a]; Ali Atharifard[a]; James Atlas[b]; Stephen T. Bell[a]; Anthony P. H. Butler[a,b,c,d,e]; Philip H. Butler[a,b,c,d,e]; Pierre Carbonez[c,d]; Claire Chambers[a]; Krishna M. Chapagain[a,c]; Alexander I. Chernoglazov[a,e]; Jennifer A. Clark[a,c,f]; Frances Colgan[c]; Jonathan S. Crighton[c]; Shishir Dahal[c,i,j]; Jérôme Damet[c,d,k]; Theodorus Dapamede[c]; Niels J. A. de Ruiter[a,b,c,e]; Devyani Dixit[a,b]; Robert M. N. Doesburg[a]; Karen Dombroski[a]; Neryda Duncan[a]; Steven P. Gieseg[b,c,d]; Anish Gopinathan[a]; Brian P. Goulter[a]; Joseph L. Healy[a]; Luke Holmes[l]; Kevin Jonker[a,c]; Tracy Kirkbride[f]; Chiara Lowe[a]; V. B. H. Mandalika[a,e]; Aysouda Matanaghi[a]; Mahdieh Moghiseh[a,c]; Manoj Paladugu[b]; David Palmer[g]; Raj K. Panta[a,d]; Peter Renaud[b,c]; Yann Sayous[b]; Nanette Schleich[h]; Emily Searle[b]; Jereena S. Sheeja[a]; Aaron Smith[b]; Leza Vanden Broeke[a]; V. S. Vivek[a]; E. Peter Walker[c]; Manoj Wijesooriya[a]; W. Ross Younger[a]

[a]MARS Bioimaging Limited, Christchurch, New Zealand
[b]University of Canterbury, Christchurch, New Zealand
[c]University of Otago Christchurch, Christchurch, New Zealand
[d]European Organisation for Nuclear Research (CERN), Geneva, Switzerland
[e]Human Interface Technology Laboratory New Zealand, University of Canterbury, Christchurch, New Zealand
[f]Ara Institute of Canterbury, Christchurch, New Zealand
[g]Lincoln University, Lincoln, New Zealand
[h]University of Otago Wellington, Wellington, New Zealand
[i]Ministry of Health, Kathmandu, Nepal
[j]National Academy of Medical Sciences, Kathmandu, Nepal
[k]Institute of Radiation Physics, Lausanne University Hospital, Lausanne, Switzerland
[l]Canterbury District Health Board, Christchurch, New Zealand

References

1. Smith L, Byrne HL, Waddington D, et al. Nanoparticles for MRI-guided radiation therapy: a review. Cancer. Nano. 2022;13(38) https://doi.org/10.1186/s12645-022-00145-8.
2. Thakor AS, et al. Clinically approved nanoparticle imaging agents. Journal of nuclear medicine: official publication, Society of Nuclear Medicine. 2016;57(12):1833–7. https://doi.org/10.2967/jnumed.116.181362.
3. Zlitni A, Gambhir SS. Molecular imaging agents for ultrasound. Curr Opin Chem Biol. 2018;45:113–20.
4. Gaikwad HK, et al. Molecular imaging of cancer using X-ray computed tomography with protease targeted iodinated activity-based probes. Nano Lett. 2018;18(3):1582–91.
5. Li H, Meade TJ. Molecular magnetic resonance imaging with Gd (III)-based contrast agents: challenges and key advances. J Am Chem Soc. 2019;141(43):17025–41.
6. Wu M, Shu J. Multimodal molecular imaging: current status and future directions. Contrast Media Mol Imaging. 2018;2018:1.
7. Hussain T, Nguyen QT. Molecular imaging for cancer diagnosis and surgery. Adv Drug Deliv Rev. 2014;66:90–100.
8. Terreno E, et al. Challenges for molecular magnetic resonance imaging. Chem Rev. 2010;110(5):3019–42.

9. MacRitchie N, et al. Molecular imaging of inflammation-current and emerging technologies for diagnosis and treatment. Pharmacol Ther. 2020;211:107550.

10. Toczek J, et al. Computed tomography imaging of macrophage phagocytic activity in abdominal aortic aneurysm. Theranostics. 2021;11(12):5876.

11. Lu F-M, Yuan Z. PET/SPECT molecular imaging in clinical neuroscience: recent advances in the investigation of CNS diseases. Quant Imaging Med Surg. 2015;5(3):433.

12. Borga M, et al. Brown adipose tissue in humans: detection and functional analysis using PET (positron emission tomography), MRI (magnetic resonance imaging), and DECT (dual energy computed tomography). Methods Enzymol. 2014;537:141–59.

13. Polycarpou I, et al. Impact of respiratory motion correction and spatial resolution on lesion detection in PET: a simulation study based on real MR dynamic data. Phys Med Biol. 2014;59(3):697.

14. Jansen FP, Vanderheyden J-L. The future of SPECT in a time of PET. Nucl Med Biol. 2007;34(7):733–5.

15. Kaushik A, et al. Estimation of radiation dose to patients from 18FDG whole body PET/CT investigations using dynamic PET scan protocol. Indian J Med Res. 2015;142(6):721.

16. von Schulthess GK, Pelc NJ. Integrated-modality imaging: the best of both worlds. Acad Radiol. 2002;9(11):1241–4.

17. Zaidi H, Montandon M-L, Alavi A. The clinical role of fusion imaging using PET, CT, and MR imaging. PET Clin. 2008;3(3):275–91.

18. Clark DP, Badea C. Micro-CT of rodents: state-of-the-art and future perspectives. Phys Med. 2014;30(6):619–34.

19. Willemink MJ, et al. Photon-counting CT: technical principles and clinical prospects. Radiology. 2018;289(2):293–312.

20. Shikhaliev PM, Fritz SG. Photon counting spectral CT versus conventional CT: comparative evaluation for breast imaging application. Phys Med Biol. 2011;56(7):1905–30.

21. Hsieh SS, et al. Photon counting CT: clinical applications and future developments. IEEE Trans Radiat Plasma Med Sci. 2020;5(4):441–52.

22. Vanden Broeke L, et al. Feasibility of photon-counting spectral CT in dental applications—a comparative qualitative analysis. BDJ Open. 2021;7(1):1–8.

23. Stamp LK, et al. Clinical utility of multi-energy spectral photon-counting computed tomography in crystal arthritis. Arthritis Rheumatol. 2019;71(7):1158–62.

24. Searle EK, et al. Distinguishing iron and calcium using MARS spectral CT. In: 2018 IEEE nuclear science symposium and medical imaging conference proceedings (NSS/MIC). IEEE; 2018.

25. Moghiseh M, et al. Spectral photon-counting molecular imaging for quantification of monoclonal antibody-conjugated gold nanoparticles targeted to lymphoma and breast cancer: an in vitro study. Contrast Media Mol Imaging. 2018;2018:1.

26. Chakaya J, et al. Global Tuberculosis Report 2020–Reflections on the Global TB burden, treatment and prevention efforts. Int J Infect Dis. 2021;113:S7–S12.

27. Dye C. Global epidemiology of tuberculosis. Lancet. 2006;367(9514):938–40.

28. Shah NS, et al. Extensively drug-resistant tuberculosis in the United States, 1993–2007. JAMA. 2008;300(18):2153–60.

29. Willmann R, et al. Improving translatability of preclinical studies for neuromuscular disorders: lessons from the TREAT-NMD Advisory Committee for Therapeutics (TACT). Dis Model Mech. 2020;13(2):dmm042903.

30. Hisham AN, Yip C-H. Overview of breast cancer in Malaysian women: a problem with late diagnosis. Asian J Surg. 2004;27(2):130–3.

31. Walter FM, et al. Symptoms and other factors associated with time to diagnosis and stage of lung cancer: a prospective cohort study. Br J Cancer. 2015;112(1):6–13.

32. Doubeni CA, et al. Screening colonoscopy and risk for incident late-stage colorectal cancer diagnosis in average-risk adults. Ann Intern Med. 2013;158(5):312–21.

33. Dizon DS, et al. Clinical cancer advances 2016: annual report on progress against cancer from the American Society of Clinical Oncology. J Clin Oncol. 2016;34(9):987–1011.

34. Ben-Haim S, Ell P. 18F-FDG PET and PET/CT in the evaluation of cancer treatment response. J Nucl Med. 2009;50(1):88–99.
35. Cormode DP, et al. Multicolor spectral photon-counting computed tomography: in vivo dual contrast imaging with a high count rate scanner. Sci Rep. 2017;7(1):4784.
36. Groheux D, et al. Correlation of high 18 F-FDG uptake to clinical, pathological and biological prognostic factors in breast cancer. Eur J Nucl Med Mol Imaging. 2011;38(3):426–35.
37. Liu Y, Zhai X, Wu Y. Biological correlation between glucose transporters, Ki-67 and 2-deoxy-2-[18F]-fluoro-D-glucose uptake in diffuse large B-cell lymphoma and natural killer/T-cell lymphoma. Genet Mol Res. 2016;15(2).
38. Zhong J, Di L, Zheng W. Synchronous breast cancer and breast lymphoma: two case reports and literature review. Chin J Cancer Res. 2014;26(3):355.
39. Lusic H, Grinstaff MW. X-ray-computed tomography contrast agents. Chem Rev. 2013;113(3):1641–66.
40. Moghiseh M, et al. Identification and quantification of multiple high-Z materials by spectral CT. European Congress of Radiology-ECR 2017; 2017.
41. Nam J, et al. Surface engineering of inorganic nanoparticles for imaging and therapy. Adv Drug Deliv Rev. 2013;65(5):622–48.
42. Aamir R. Using MARS spectral CT for identifying biomedical nanoparticles. Christchurch: Department of Physics & Astronomy, University of Canterbury; 2013.
43. Li C-Y, et al. Initial stages of tumor cell-induced angiogenesis: evaluation via skin window chambers in rodent models. J Natl Cancer Inst. 2000;92(2):143–7.
44. Kumagai Y, et al. Tumor-associated macrophages and angiogenesis in early stage esophageal squamous cell carcinoma. Esophagus. 2016;13(3):245–53.
45. Ribatti D, Pezzella F. Overview on the different patterns of tumor vascularization. Cell. 2021;10(3):639.
46. Márquez-Garbán DC, et al. Squalamine blocks tumor-associated angiogenesis and growth of human breast cancer cells with or without HER-2/neu overexpression. Cancer Lett. 2019;449:66–75.
47. Pietras RJ. Interactions between estrogen and growth factor receptors in human breast cancers and the tumor-associated vasculature. Breast J. 2003;9(5):361–73.
48. Benazzi C, et al. Angiogenesis in spontaneous tumors and implications for comparative tumor biology. Sci World J. 2014;2014:1.
49. Soliman MA, et al. Current concepts in multi-modality imaging of solid tumor angiogenesis. Cancers. 2020;12(11):3239.
50. Turkbey B, et al. Imaging of tumor angiogenesis: functional or targeted? Am J Roentgenol. 2009;193(2):304–13.
51. Amma MR, et al. Assessment of metal implant induced artefacts using photon counting spectral CT. In: Developments in X-ray tomography XII. International Society for Optics and Photonics; 2019.
52. Moghiseh M, et al. Cancer imaging with nanoparticles using MARS spectral scanner. In: 2018 IEEE nuclear science symposium and medical imaging conference proceedings (NSS/MIC). IEEE; 2018.
53. Moghiseh M. Optimization of spectral CT for novel applications of nanoparticles. In: Radiology. Christchurch: Otago; 2018.
54. Moghiseh M, et al. Spectral CT of cancer cells with nanoparticles: in vitro and in vivo results. In: European congress of radiology. Vienna: European Society of Radiology (ESR); 2019.
55. Benjamin EJ, et al. Heart disease and stroke statistics-2018 update: a report from the American Heart Association. Circulation. 2018;137(12):e67.
56. Frohlich J, Al-Sarraf A. Cardiovascular risk and atherosclerosis prevention. Cardiovasc Pathol. 2013;22(1):16–8.
57. Obaid DR, et al. Coronary CT angiography features of ruptured and high-risk atherosclerotic plaques: correlation with intra-vascular ultrasound. J Cardiovasc Comput Tomogr. 2017;11(6):455–61.

58. Kolossváry M, et al. Plaque imaging with CT—a comprehensive review on coronary CT angiography based risk assessment. Cardiovasc Diagn Ther. 2017;7(5):489.
59. Dweck MR, et al. MR imaging of coronary arteries and plaques. JACC Cardiovasc Imaging. 2016;9(3):306–16.
60. Bom MJ, et al. Early detection and treatment of the vulnerable coronary plaque: can we prevent acute coronary syndromes? Circ Cardiovasc Imaging. 2017;10(5):e005973.
61. Tarkin JM, et al. Imaging atherosclerosis. Circ Res. 2016;118(4):750–69.
62. Ding Y, et al. Gold nanoparticles for nucleic acid delivery. Mol Ther. 2014;22(6):1075–83.
63. Chen IY, Wu JC. Cardiovascular molecular imaging: focus on clinical translation. Circulation. 2011;123(4):425–43.
64. Zainon R, et al. Spectral CT of carotid atherosclerotic plaque: comparison with histology. Eur Radiol. 2012;22(12):2581–8.
65. Wickline SA, et al. Molecular imaging and therapy of atherosclerosis with targeted nanoparticles. J Magn Reson Imaging. 2007;25(4):667–80.
66. Chhour P, et al. Labeling monocytes with gold nanoparticles to track their recruitment in atherosclerosis with computed tomography. Biomaterials. 2016;87:93–103.
67. Ferguson C, et al. Does adult fracture repair recapitulate embryonic skeletal formation? Mech Dev. 1999;87(1–2):57–66.
68. Marsell R, Einhorn TA. The biology of fracture healing. Injury. 2011;42(6):551–5.
69. Kon T, et al. Expression of osteoprotegerin, receptor activator of NF-κB ligand (osteoprotegerin ligand) and related proinflammatory cytokines during fracture healing. J Bone Miner Res. 2001;16(6):1004–14.
70. Street J, et al. Vascular endothelial growth factor stimulates bone repair by promoting angiogenesis and bone turnover. Proc Natl Acad Sci. 2002;99(15):9656–61.
71. Gerber H-P, Ferrara N. Angiogenesis and bone growth. Trends Cardiovasc Med. 2000;10(5):223–8.
72. Lehmann W, et al. Tumor necrosis factor alpha (TNF-α) coordinately regulates the expression of specific matrix metalloproteinases (MMPS) and angiogenic factors during fracture healing. Bone. 2005;36(2):300–10.
73. Kranioti EF, Bonicelli A, García-Donas JG. Bone-mineral density: clinical significance, methods of quantification and forensic applications. Res Rep Forensic Med Sci. 2019;9:9–21.
74. Mazess R, et al. Advances in noninvasive bone measurement. Ann Biomed Eng. 1989;17(2):177–81.
75. Syed Z, Khan A. Bone densitometry: applications and limitations. J Obstet Gynaecol Can. 2002;24(6):476–84.
76. Griffith JF, Genant HK. New imaging modalities in bone. Curr Rheumatol Rep. 2011;13(3):241–50.
77. Brismar TB, Budinsky L, Majumdar S. Evaluation of trabecular bone orientation in wrists of young volunteers using MR relaxometry and high resolution MRI. In: Noninvasive assessment of trabecular bone architecture and the competence of bone. Springer; 2001. p. 1–7.
78. Ammann P, Rizzoli R. Bone strength and its determinants. Osteoporos Int. 2003;14(3):13–8.
79. Judex S, et al. Combining high-resolution micro-computed tomography with material composition to define the quality of bone tissue. Curr Osteoporos Rep. 2003;1(1):11–9.
80. Pothuaud L, et al. Combination of topological parameters and bone volume fraction better predicts the mechanical properties of trabecular bone. J Biomech. 2002;35(8):1091–9.
81. Genant H, Engelke K, Prevrhal S. Advanced CT bone imaging in osteoporosis. Rheumatology. 2008;47(suppl_4):iv9–iv16.
82. Stagi S, et al. Peripheral quantitative computed tomography (pQCT) for the assessment of bone strength in most of bone affecting conditions in developmental age: a review. Ital J Pediatr. 2016;42(1):1–20.
83. Gebauer M, et al. DXA and pQCT predict pertrochanteric and not femoral neck fracture load in a human side-impact fracture model. J Orthop Res. 2014;32(1):31–8.

84. Boutroy S, et al. In vivo assessment of trabecular bone microarchitecture by high-resolution peripheral quantitative computed tomography. J Clin Endocrinol Metabol. 2005;90(12):6508–15.
85. Rüegsegger P, Koller B, Müller R. A microtomographic system for the nondestructive evaluation of bone architecture. Calcif Tissue Int. 1996;58(1):24–9.
86. Kazakia GJ, et al. In vivo determination of bone structure in postmenopausal women: a comparison of HR-pQCT and high-field MR imaging. J Bone Miner Res. 2008;23(4):463–74.
87. Naseri Kouzehgarani G. Relationship between diabetes and bone health status in adults with diabetes. 2012.
88. Dhainaut A, et al. Technologies for assessment of bone reflecting bone strength and bone mineral density in elderly women: an update. Womens Health. 2016;12(2):209–16.
89. Chin K-Y, Ima-Nirwana S. Calcaneal quantitative ultrasound as a determinant of bone health status: what properties of bone does it reflect? Int J Med Sci. 2013;10(12):1778.
90. Mesquita AQD, Barbieri G, Barbieri CH. Correlation between ultrasound velocity and densitometry in fresh and demineralized cortical bone. Clinics. 2016;71:657–63.
91. Rajeswari Amma M. Study of bone-metal interface in orthopaedic application using spectral CT. University of Otago; 2020.
92. Ramyar M. MARS spectral CT technology for simultaneous assessment of articular cartilage and bone. In: Radiology. University of Otago; 2017.
93. Ostadhossein F, et al. Multi-"color" delineation of bone microdamages using ligand-directed sub-5 nm hafnia nanodots and photon counting CT imaging. Adv Funct Mater. 2020;30(4):1904936.
94. Chapurlat R, Delmas P. Bone microdamage: a clinical perspective. Osteoporos Int. 2009;20(8):1299–308.
95. O'Brien FJ, et al. Microcracks in cortical bone: how do they affect bone biology? Curr Osteoporos Rep. 2005;3(2):39–45.
96. Acevedo C, et al. Fatigue as the missing link between bone fragility and fracture. Nat Biomed Eng. 2018;2(2):62–71.
97. Odvina CV, et al. Unusual mid-shaft fractures during long-term bisphosphonate therapy. Clin Endocrinol. 2010;72(2):161–8.
98. Brandi ML. Microarchitecture, the key to bone quality. Rheumatology. 2009;48(suppl_4):iv3–8.
99. Maldonado M, Nam J. The role of changes in extracellular matrix of cartilage in the presence of inflammation on the pathology of osteoarthritis. BioMed Res Int. 2013;2013.
100. Bansal PN, et al. Cationic contrast agents improve quantification of glycosaminoglycan (GAG) content by contrast enhanced CT imaging of cartilage. J Orthop Res. 2011;29(5):704–9.
101. Kubaski F, et al. Glycosaminoglycans detection methods: applications of mass spectrometry. Mol Genet Metab. 2017;120(1–2):67–77.
102. Martel-Pelletier J, et al. Cartilage in normal and osteoarthritis conditions. Best Pract Res Clin Rheumatol. 2008;22(2):351–84.
103. Buckwalter JA, Mow VC, Ratcliffe A. Restoration of injured or degenerated articular cartilage. JAAOS-J Am Acad Orthop Surg. 1994;2(4):192–201.
104. Frenkel S, et al. Regeneration of articular cartilage–evaluation of osteochondral defect repair in the rabbit using multiphasic implants. Osteoarthr Cartil. 2005;13(9):798–807.
105. Kumar P, Clark ML. Kumar & Clark's cases in clinical medicine e-book. Elsevier Health Sciences; 2020.
106. Benders K, et al. Formalin fixation affects equilibrium partitioning of an ionic contrast agent-microcomputed tomography (EPIC-μCT) imaging of osteochondral samples. Osteoarthr Cartil. 2010;18(12):1586–91.
107. Farndale RW, Buttle DJ, Barrett AJ. Improved quantitation and discrimination of sulphated glycosaminoglycans by use of dimethylmethylene blue. Biochim Biophys Acta Gen Subj. 1986;883(2):173–7.

108. Gheno R, et al. Musculoskeletal disorders in the elderly. J Clin Imaging Sci. 2012;2.
109. Douka M, et al. Imaging of shoulder arthropathies. Hell J Radiol. 2021;6(1).
110. Disler DG, Recht MP, McCauley TR. MR imaging of articular cartilage. Skelet Radiol. 2000;29(7):367–77.
111. Stubendorff J, et al. Is cartilage sGAG content related to early changes in cartilage disease? Implications for interpretation of dGEMRIC. Osteoarthr Cartil. 2012;20(5):396–404.
112. Saxena V, et al. T1ρ magnetic resonance imaging to assess cartilage damage after primary shoulder dislocation. Am J Sports Med. 2016;44(11):2800–6.
113. Mosher TJ, Liu Y, Torok CM. Functional cartilage MRI T2 mapping: evaluating the effect of age and training on knee cartilage response to running. Osteoarthr Cartil. 2010;18(3):358–64.
114. Zbýň Š, et al. Sodium MR imaging of articular cartilage pathologies. Curr Radiol Rep. 2014;2(4):41.
115. Palmer AW, Guldberg RE, Levenston ME. Analysis of cartilage matrix fixed charge density and three-dimensional morphology via contrast-enhanced microcomputed tomography. Proc Natl Acad Sci. 2006;103(51):19255–60.
116. Xie L, et al. Quantitative assessment of articular cartilage morphology via EPIC-μCT. Osteoarthr Cartil. 2009;17(3):313–20.
117. Entezari V. Cationic contrast agents improve quantification of glycosaminoglycan (GAG) content by contrast enhanced CT imaging of cartilage. 2010.
118. Baer K, et al. Spectral CT imaging of human osteoarthritic cartilage via quantitative assessment of glycosaminoglycan content using multiple contrast agents. APL Bioeng. 2021;5(2):026101.
119. Rajendran K, et al. Quantitative imaging of excised osteoarthritic cartilage using spectral CT. Eur Radiol. 2017;27(1):384–92.
120. Moghiseh M, et al. Discrimination of multiple high-Z materials by multi-energy spectral CT– A phantom study. JSM Biomed Imaging Data Pap. 2016;61:1007.
121. Nestorova R, Fodor D. Crystal-induced arthritis. In: Musculoskeletal ultrasonography in rheumatic diseases. Springer; 2015. p. 137–67.
122. Neogi T, et al. 2015 gout classification criteria: an American College of Rheumatology/European League Against Rheumatism collaborative initiative. Arthritis Rheumatol. 2015;67(10):2557–68.
123. Zhang W, et al. European league against rheumatism recommendations for calcium pyrophosphate deposition. Part I: terminology and diagnosis. Ann Rheum Dis. 2011;70(4):563–70.
124. Becce F, et al. Winds of change in imaging of calcium crystal deposition diseases. Joint Bone Spine. 2019;86(6):665–8.
125. Freire V, Moser TP, Lepage-Saucier M. Radiological identification and analysis of soft tissue musculoskeletal calcifications. Insights Imaging. 2018;9(4):477–92.
126. Filippou G, et al. Ultrasound in the diagnosis of calcium pyrophosphate dihydrate deposition disease. A systematic literature review and a meta-analysis. Osteoarthr Cartil. 2016;24(6):973–81.
127. Bernabei I, et al. Multi-energy photon-counting computed tomography versus other clinical imaging techniques for the identification of articular calcium crystal deposition. Rheumatology. 2021;60(5):2483–5.
128. Berger MJ, Hubbell JH, Seltzer SM, Chang J, Coursey JS, Sukumar R, Zucker DS, Olsen K. XCOM: photon cross section database. 2010.
129. Lowe C., et al. Molecular imaging of pulmonary tuberculosis in an ex-vivo mouse model using spectral photon-counting computed tomography and micro-CT. IEEE. 2021.

Quantitative Breast Lesion Characterization with Spectral Mammography: A Feasibility Study

Huanjun Ding and Sabee Molloi

1 Introduction

Currently, the standard imaging modality for breast cancer screening is mammography [1–5]. Yet, despite mammography's impressive advantages in detection performance, imaging time, and cost-effectiveness, its limitations are widely recognized [6]. One of the biggest challenges for accurate early detection in breast cancer screening is increasing the specificity of tumor detection to avoid recalling healthy women and exposing them to follow-up examinations that may involve additional radiation or a needle biopsy, which are stressful to the patient and costly. With the conventional mammograms, the radiologist makes the determination based on specific features in mammographic patterns such as shape, size, margin, or pattern of abnormal density [7, 8]. However, in contrast to stellate lesions, the most suspicious findings of circular and oval lesions are relatively easy to detect, but difficult to characterize as benign or malignant, and therefore require follow-up examination.

A recent report suggests that, in addition to irregular mass shape, speculated mass margin, and patient age, a high mammographic attenuation of a mass increases its likelihood of malignancy [9]. However, high mass density by itself is not sufficiently accurate to avert the need for a biopsy. If lesions could be characterized quantitatively according to their chemical compositions, predictive capability might be improved. Thus, there is increased interest in developing lesion characterization techniques that can characterize circular lesions as benign or malignant during the initial screening. It has been suggested that malignant tumors have reduced lipid and increased water contents compared to normal breast tissue [10–13]. In addition, other reports suggest a positive correlation between increased tissue water content

H. Ding · S. Molloi (✉)
Department of Radiological Sciences, University of California, Irvine, CA, USA
e-mail: symolloi@hs.uci.edu

© The Author(s), under exclusive license to Springer Nature Switzerland AG 2023
S. Hsieh, K. (Kris) Iniewski (eds.), *Photon Counting Computed Tomography*,
https://doi.org/10.1007/978-3-031-26062-9_5

and carcinogenesis [14]. X-ray linear attenuation coefficient of breast lesions has been studied in the past and was reported to be significantly different from fibro-glandular tissue [15, 16]. A recent study also suggested that the attenuation of cyst fluid was found to be significantly different from that of water [17]. These reports indicate that mammography's sensitivity and specificity may be improved if breast tissue composition can be accurately measured to improve characterization of lesions according to their composition.

Dual-energy imaging can exploit differences between specific types of breast tissues due to their unique effective atomic numbers (Z), providing separate quantitative thickness measurements for each tissue. This technique has been successfully implemented for breast density quantification by using the standard screening mammogram such as the low-energy image and an additional low-dose high-energy image [18–20]. However, the high-energy image may slightly increase the mean glandular dose and result in misregistration artifacts due to patient motion. These potential challenges have been successfully addressed with the recent developments in photon-counting X-ray detectors. A spectral mammography system based on photon-counting detectors in a scanning multi-slit geometry [21] allows dual-energy data acquisition to be completed with a single exposure. A user-defined energy threshold is used to sort photons into low- and high-energy bins, according to their energies. Compared to traditional dual-kVp technique, energy-resolved photon-counting detectors minimize the spectral overlap and completely eliminate the misregistration artifacts in dual-energy decomposition. In addition, conventional charge-integrating detectors generally work in the current mode, which integrates both the signal and noise from the detector and electronics over time. The presence of electronic noise can substantially reduce the signal-to-noise ratio (SNR) in low-dose imaging [22]. On the other hand, photon-counting detectors can eliminate electronic noise, which offers promising potential for low-dose imaging [23]. A previous report shows that an ideal photon-counting detector with good energy resolution outperforms conventional charge-integrating detectors in image quality and various detection tasks [24]. It has been shown that the photon-counting spectral mammography system can accurately measure glandular and adipose tissue thicknesses for breast density quantification [25]. Thus, it is of interest to investigate the feasibility of characterizing breast lesion composition using dual-energy decomposition of spectral mammography images acquired on a photon-counting system.

The fundamental chemical components in either normal or cancerous breast tissues are water, lipid, and protein. A previous report has proposed to characterize breast tissue using a three-material compositional measure of water, lipid, and protein contents with dual-energy mammography [26]. The study suggested that knowledge of the composition of breast lesions and their periphery appeared additive in combination with existing diagnostic methods for the distinction between different benign and malignant lesion types [26]. However, the results failed to reach statistical significance. One potential issue of using a three-component model with dual-energy mammography is that it requires accurate breast thickness measurement as the third independent physical measurement. Unfortunately, this is difficult to obtain in practice, particularly in the periphery of the breast where the breast is not in contact with the compression plate. Uncertainties in thickness estimation, induced by the shape model and the mechanical precision of the compression paddle, can lead to significant errors in lesion compositional analysis, which can reduce the predictive power for malignancy. To address this potential issue, we propose to characterize breast lesions with a two-compartment model using water and lipid as the basis material. Our previous postmortem study has suggested that protein generally contributes less than 6% in breast tissue [27]. More importantly, using a two-compartment model, protein in breast tissue will be decomposed into water and lipid basis materials. Since the dual-energy decomposition coefficients are well defined for the two basis materials, the presence of a small amount of protein will only add a small systematic offset in estimation of water and lipid contents. This approach has been tested in a simulation study, which suggested that the measured water content using a two-compartment model correlated strongly with the known values of water in breast tissue [28]. In this chapter, we investigate the feasibility of characterizing breast lesions using a two-compartment model with physical phantoms and tissue samples. The purpose of this study is to evaluate the accuracy and precision of the proposed technique for lesion characterization.

2 Materials and Methods

2.1 Spectral Mammography System

A spectral mammography system (MircoDose SI L50, Philips Inc.) was used during the studies, which is able to acquire dual-energy images within a single exposure. The system consists of a tungsten-anode X-ray tube, an Al filter, a pre- and a post-collimator, and a Si-strip photon-counting detector unit, which are all mounted on a common arm that can rotate around the center of the source, allowing the collimators and the photon counting detector to scan relative to the compressed breast. The detector's energy resolution at the mammography energy range is approximately 5 keV at full width half maximum (FWHM) [29]. The electronic readout noise is effectively eliminated by selecting an appropriate low-energy threshold. A multi-slit collimator shapes the beam to match the detector, and a two-dimensional image is generated when the beam and detector are scanned relative to the breast. The scanning multi-slit technique helps to eliminate scattered radiation, which further improves the SNR. A previous study suggests that the scatter to primary ratio (SPR) for this geometry is expected to be less than 6% for phantom thicknesses ranging from 3 to 7 cm at various tube voltages [30]. More details about the spectral mammography system can be found in a previous publication [25].

2.2   Dual-Energy Calibration

A calibration phantom consisting of two compartments, made from plastic water[®] LR and adipose equivalent plastics (CIRS, Norfolk, VA), was constructed for dual-energy decomposition. The X-ray attenuation coefficients of plastic water were calculated based on its elemental compositions provided by the manufacturer and were found to closely resemble those of pure liquid water in the mammographic energy range, which provides a reason for the use of solid water as a substitute for liquid water in the dual-energy calibration. Unfortunately, it is difficult to fabricate a plastic-based lipid phantom that can represent real lipid. Instead, we used the adipose equivalent phantom, which consists of approximately 15% water and 85% lipid, according to its chemical composition [31]. After dual-energy decomposition by using plastic water and adipose as the basis materials, a linear transformation can be applied to convert the basis material into water and lipid based on the calibrated composition of the adipose equivalent phantom [17]. A schematic drawing of the calibration phantom is shown in Fig. 1. It had a stair shape with four different heights: 2, 4, 6, and 8 cm, where each of the heights had five different plastic water densities ranging from 0% to 100% with an increasing step of 25%. This design provided a total of 20 calibration points that cover both thickness and density variations. The calibration phantom was carefully machined and the thickness of each step was confirmed with caliper measurements, which suggested an uncertainty of approximately 25 µm. The calibration phantom was imaged at all the available tube voltages on the system, including 26, 29, 32, 35, and 38 kVp. The spectral mammography system has a built-in calibration for the selection of the high-energy threshold, which depends on the tube voltage and phantom thickness. The log signals from the low- and high-energy images were derived for all 20 calibration points.

In dual-energy material decomposition, the low- and high-energy signals of a tissue with given thickness can be written as a linear combination of the two basis materials used in the calibration [18]. The two measurements from different energy bins provide enough information to solve the thickness of the two basis

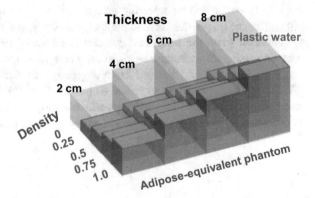

Fig. 1 A schematic drawing of the calibration phantom composed of plastic water and adipose equivalent material of various thicknesses and densities

materials. Due to the poly-energetic nature of the diagnostic X-ray spectra, dual-energy imaging usually uses higher-order inverse functions that map the log signals into content thicknesses. In this study, a nonlinear rational fitting function was used, which has been shown to have high fitting accuracy [32], to calibrate the imaging system:

$$t = \frac{a_0 + a_1 S_L + a_2 S_H + a_3 S_L{}^2 + a_4 S_H S_L + a_5 S_H{}^2 + a_6 S_L{}^3 + a_7 S_H{}^3}{1 + b_1 S_L + b_2 S_H + b_3 S_L{}^2 + b_4 S_H{}^2} \tag{1}$$

The calibration process will substitute the known thickness values, t_i, for either plastic water or adipose equivalent phantoms, and the corresponding dual-energy attenuation measurements S_L and S_H into Eq. (1). Subsequently, the system calibration parameters (a_0, a_1 . . .) for each material will be determined separately from a nonlinear least-squares minimization algorithm [33]. Using the measured dual-energy log-signals, the calibration parameters will then be used as the decomposition coefficients to characterize lesions in terms of water and lipid contents. This method has been previously validated in dual-energy breast-density quantification studies, which have shown good accuracy for tissue compositional analysis [25].

2.3   Phantom and Tissue Validation

To validate the proposed method, phantom studies were designed using a mammographic phantom with a heterogeneous pattern (BR3D, CIRS, Norfolk, VA) as the background in order to simulate the anatomical noise in clinical mammograms. The BR3D phantom is constructed with a swirled pattern of adipose and glandular tissue equivalent materials with an overall 50–50 ratio. The random swirling pattern makes the background different depending on the position on the phantom, which leads to variations of the local glandularity. The lesion phantoms were made from plastic water and adipose disks that are 1 mm in thickness and 2 cm in diameter. Five of those disks were stacked together in order to simulate a 0.5 cm lesion. The configuration of the lesion disk phantoms was varied to produce six possible plastic water densities, which ranged from 0% to 100% with an increasing step of 20%. To embed the lesion disk phantoms in the breast background, a 0.5 cm thick adipose phantom containing two holes 2 cm in diameter were used to hold the lesion disks. The adipose phantom with the embedded lesion phantoms was then placed on top of the BR3D phantoms, which created a breast phantom that has a uniform thickness around the lesions. This phantom design approximately resembles a breast under compression in clinical mammography scans.

Four phantom validation studies were designed, which addressed the effects of lesion composition, breast thickness, breast density, and lesion locations. The first study imaged the lesion phantoms with 4 cm thick BR3D background. Lesions of various water densities ranging from 0% to 100% were imaged at different locations on the BR3D background. The second study fixed the water density of the lesion phantom at 40% and imaged it on BR3D phantoms of different thicknesses ranging from 2 to 5 cm. The third study used the same 40% density lesion phantom, but imaged it with a series of glandular and adipose tissue equivalent phantoms with a total thickness of 4 cm. The combinations of the two background phantoms were varied to provide a set of breast backgrounds of different densities ranging from 0% to 100%. Finally, in the last phantom study, the lesions were inserted in different levels of a 4 cm BR3D background, which simulated different locations of a lesion in the projection direction.

To further test the lesion characterization method in tissue, an in vitro study was designed using bovine tissue as the background. Pure lean and fat bovine tissue pieces were mixed together to form heterogeneous background phantoms of 25%, 50%, and 80% lean weight percentage. The mixture bovine tissue samples were kept in thin, plastic bags and shaped to resemble compressed breasts. The previously discussed lesion disk phantoms of various water densities were placed inside the tissue background and imaged at different locations. In addition, to better simulate the lesions, lesion phantoms were also made from cut lean and fat bovine tissues. The lean and fatty tissues were carefully weighed to produce various lesion phantoms with lean mass percentage ranging from 0% to 100% with an increasing step of 10%. The tissue lesion phantoms were first wrapped with a thin transparent plastic and then placed in a plastic tube with a diameter and height of 10 mm. This setup helped to fix the shape of the lesion phantoms and also allowed us to keep track of the locations in the tissue background. After imaging, all tissue lesion phantoms were decomposed into water, lipid, and protein using a previously reported chemical analysis technique [27]. The measured weight of water, lipid, and protein contents was converted into volumes using the corresponding density values. Volumetric densities were calculated for water as the reference standard.

Finally, postmortem breasts were also imaged with the previously mentioned lesion disk phantoms of various water densities. Lesion phantoms were embedded at different locations inside the breast tissue. Postmortem breasts were imaged using the spectral mammography system with clinical exposure settings. Depending on the compression thicknesses of different breast samples, the tube voltages ranged between 26 and 38 kVp and the current-time productions ranged between 10 and 14 mAs. The average glandular dose reported from the system ranged from 0.2 to 0.7 mGy, which is significantly less than the clinical standard for single-view mammography.

2.4 Image Processing

All images in the validation study were acquired using the automatic exposure setting and the recommended tube voltage from the built-in screening protocol for clinical applications. Dual-energy images in the raw format generated from the spectral mammography system were first normalized by the corresponding open field images at the same tube voltage settings. Then the log signals were calculated for both low- and high-energy images and were used for material decomposition using Eq. (1) with the calibrated decomposition coefficients for plastic water and adipose. The decomposition process converted the low- and high-energy log signal images into thickness images of the two basis materials, where the pixel values represent the total plastic water/adipose thickness inside a column of sample above the corresponding detector pixel. In the case where a lesion phantom is inside this volume, the decomposed plastic water signal can then be written as the sum of the plastic water thickness in the lesion and in the background above and below the lesion. The lesion's plastic water content can then be isolated by subtracting the total decomposed signal from a region outside the lesion, under the assumption that the plastic water content distribution in the background is locally uniform in a region that has a similar size as the lesion. Thus, the subtracted signals can be used to characterize lesions with different chemical compositions in terms of plastic water and adipose contents.

To measure the signal difference between the lesion and the surrounding area, two ROIs were manually delineated on the dual-energy decomposed images. A circular ROI was used to measure the decomposed material thicknesses inside the disks and an annular ROI was used to measure the thickness signal from an adjacent area near the lesion phantoms. The two ROIs were separated by a narrow region with a thickness of approximately 1 mm in order to avoid any potential blurring induced by magnification. A schematic illustration of the ROI selections is shown in Fig. 2. The signal difference for plastic water thickness measured on dual-energy decomposed image was compared to the known values for all lesion phantoms. Linear regression analysis was performed for all the validation studies. Root-mean-square (RMS) error was used to characterize the accuracy of the measurements. In the tissue lesion analysis, an additional step was taken to convert the measured plastic water and adipose phantom thicknesses into water and lipid thicknesses. Linear regression analysis was performed to investigate the correlation of water densities obtained through spectral mammography and chemical analysis.

3   Results

The decomposition errors from the proposed dual-energy calibration method are summarized in Table 1 for all tube voltages available with the spectral mammography system. RMS errors were calculated over all 20 calibration points, which vary in thickness and density. The RMS errors in plastic water thickness ranged between 0.11 and 0.24 mm, with the highest error of 0.24 mm at the tube voltage of 32 kVp. The adipose measurements showed a similar trend as that of the plastic water. The RMS errors in water density, which were calculated from the ratio of plastic water and total thickness, are also shown in Table 1.

Figure 2 shows an example of the images acquired during the phantom validation. In the dual-energy images (Fig. 2a, b), lesion disks of 80% density can be seen embedded within a swirled background produced by the BR3D breast phantom.

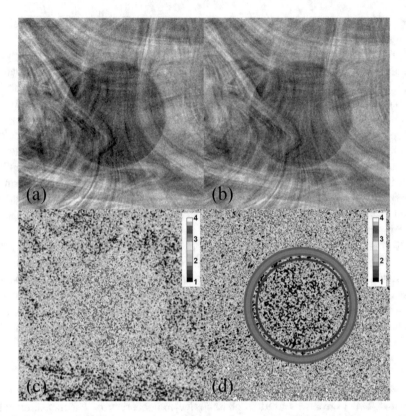

Fig. 2 (**a**) Low- and (**b**) high-energy images in the raw format for an 80% density lesion embedded in a 4 cm BR3D breast phantom, and the decomposed plastic water (**c**) and adipose (**d**) images. The colored scales are in unit of centimeter. A schematic illustration of the circular and the annular ROIs used to derive the thickness differences is shown in (**d**)

Table 1 RMS error from dual-energy calibration for all five tube voltages used in the study

	26 kVp	29 kVp	32 kVp	35 kVp	38 kVp
Plastic water (mm)	0.11	0.13	0.24	0.20	0.20
Adipose (mm)	0.22	0.2	0.33	0.25	0.11
Water density (%)	0.39	0.31	0.64	0.41	0.35

After decomposition, the thicknesses of plastic water and adipose above each pixel are presented in Fig. 2c, d, respectively. Although the current decomposition scheme will not minimize the background noise induced by the swirl pattern, its focus is to quantify the compositional thicknesses in different regions of the phantom. In a clinical application, this method will be used to measure lesion composition in the presence of anatomical noise.

In the first validation study, lesion disk phantoms of various densities were imaged at different locations on a 4 cm BR3D breast phantom, which represents

Fig. 3 Measured plastic water and adipose thicknesses in lesion phantoms of different densities as a function of the known values for a 4 cm heterogeneous BR3D phantom

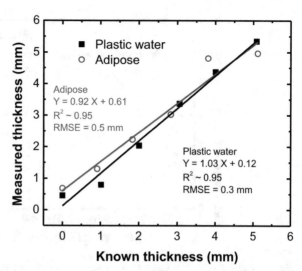

an average breast compression thickness in mammography. Plastic water thickness of the lesion measured using the method described above was compared to the known thickness of the plastic water disks in Fig. 3. The results showed a good linear correlation between the measured and the known thickness of the decomposed compartments for both plastic water and adipose. The slopes of the linear correlation lines were derived to be 1.03 and 0.92 for plastic water and adipose, respectively. The linear correlation coefficient (R^2) was determined to be 0.95 for both basis materials. The RMS errors were estimated to be 0.3 and 0.5 mm for plastic water and adipose, respectively.

Since the method to characterize lesion compositions depends on the subtraction of the normal surrounding tissue component thickness, it is necessary to investigate the effect of the background thickness and density variations on the accuracy of the thickness quantification. In Fig. 4, the measured plastic water and adipose thicknesses in a 40% water density lesion phantom are shown as a function of the background breast phantom thicknesses, which ranged from 2 to 5 cm. The expected thickness values for the two components are shown as the dashed lines in the figure. It can be seen that the errors in lesion characterization are independent of breast thickness. The RMS errors over all breast thicknesses were determined to be 0.2 and 0.3 mm for plastic water and adipose, respectively. The effect of breast density on lesion characterization accuracy is shown in Fig. 5. The same 40% density lesion was imaged with breast phantoms of various glandularities. Similar to the breast thickness study, the measured thicknesses aligned tightly around the expected values. The RMS errors were 0.1 and 0.2 mm for plastic water and adipose, respectively. This result indicated that changes in breast density do not affect the measurement accuracy for lesion characterization using the proposed method.

Due to the projection nature of mammography, the position of the lesion inside the breast may affect the magnification of the lesion. The decomposition results

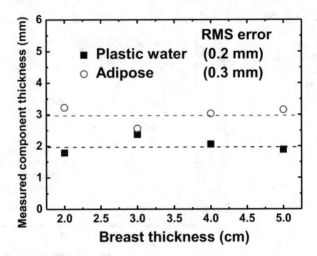

Fig. 4 Compositional measurements of a 40% density lesion in breast phantoms of different thicknesses

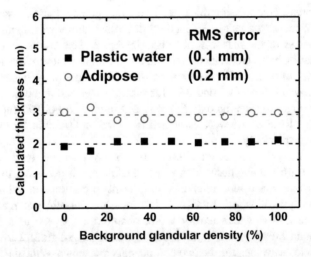

Fig. 5 Compositional measurements of a 40% density lesion in breast phantoms of different glandularity

of a 40% water density lesion imaged at different positions inside a 5 cm BR3D breast phantom is shown in Fig. 6. The calculated RMS errors in thicknesses of the decomposed plastic water and adipose were 0.2 mm and 0.3 mm, respectively. It can be seen that the accuracy of the investigated lesion characterization is not affected by the location of the lesion inside breast.

The result of lesion characterization in a tissue background is shown in Fig. 7. There is a good correlation between the calculated thickness of the decomposed materials and the known values of the disk thicknesses. In a linear regression

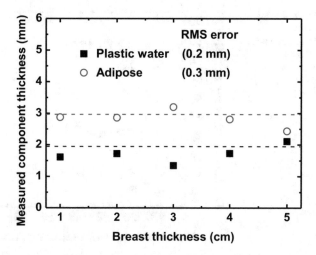

Fig. 6 Compositional measurement of a 40% lesion phantom that was positioned in different levels in 5 cm the BR3D slabs. The dashed lines indicate the known thicknesses of plastic water and adipose phantom

Fig. 7 Measured thicknesses of the decomposed plastic water and adipose as a function of the known values in lesion phantoms, which were embedded in a bovine tissue background

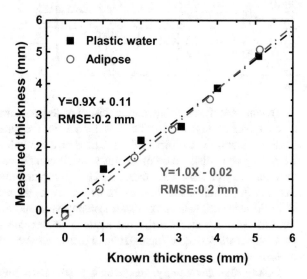

analysis, the correlation slopes for plastic water and adipose were derived to be 0.93 and 1.01, respectively. The linear correlation coefficients (R^2) for the two components were 0.98 and 0.99. RMS errors for both plastic water and adipose were estimated to be 0.2 mm. When using tissue lesion phantoms with tissue background, a good linear correlation between the decomposed water density and the reference standard from chemical analysis was observed, as shown in Fig. 8. The correlation coefficient (R^2) was estimated to be 0.96. Linear regression analysis resulted in a slope of 1.29 and an offset of −11.7%.

Fig. 8 Decomposed water density from spectral mammography as a function of that obtained from chemical analysis for meat lesions embedded in a bovine tissue background

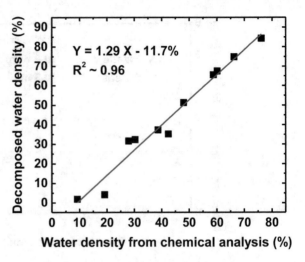

Table 2 Summary of water density measurements for lesions with various backgrounds

	Uniform	BR3D	Bovine tissue	Tissue lesion
Slope	1.02	1.03	0.93	1.29
Offset (%)	3.3	2.5	2.3	−11.7
R^2	0.99	0.95	0.98	0.96
RMS error (%)	4.2	6.1	4.9	7.4
SEE (%)	1.3	4.5	4.1	4.6

To compare the accuracy and precision of lesion characterization with different types of background, the results on water density measurements from all validation studies are summarized in Table 2. The first three columns of data were derived from the plastic disk lesion phantoms with different background, and the last set of data was derived from the bovine tissue lesion phantoms imaged with the bovine tissue background. Water density, instead of thickness, was used in the comparison, since the chemical analysis of tissue lesions phantoms can only generate volumetric information for water density calculations. Water densities of the plastic lesion phantoms were derived using the ratio of measured water thickness to the total lesion thickness.

Finally, the low-energy image of a postmortem breast with three embedded plastic lesion phantoms is shown in Fig. 9a. The plastic water densities of the three lesion phantoms were 0, 100, and 80% for locations 1, 2, and 3, respectively. The low- and high-energy images were converted into equivalent plastic water and adipose thickness images based on a technique discussed in the previous sections. As shown in Fig. 9c, two circular regions of interest or ROIs were drawn on the thickness images to calculate the thickness of the two basis materials in the lesion disks. Large donut-shaped ROIs marked by yellow concentric circles were drawn around the lesion to estimate the thickness of the basis material in the surrounding tissue. Smaller ROIs marked by a green circle were drawn within the disks to

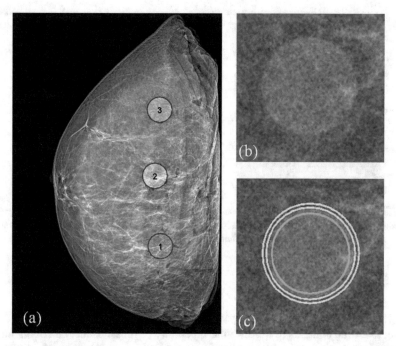

Fig. 9 Processed mammogram of a postmortem breast with three plastic lesion phantoms marked in circles (**a**). Low-energy image in the raw format for one of the lesion phantoms (**b**). A schematic illustration of the circular and the annular ROIs used to derive the thickness differences is shown in (**c**)

estimate the thicknesses of the basis materials in the lesions and the column of tissue above and below the lesions, assuming that breast tissue water density was largely spatially invariant. The measurements of thickness from the two ROIs were then subtracted to obtain the difference that corresponds to the thickness of the basis material in the lesion alone.

Figure 10a shows a plot of the average of the measured thicknesses of the plastic water in the lesion phantoms over 26 breasts against the known thicknesses. The error bars here represent the standard deviation over 26 breasts. The results demonstrate good correlation between the known and measured thicknesses with a slope of 1 and a good linear fit with R^2 of 0.99. Figure 10b is the measured plastic water density of the lesions plotted against the known density of the lesions. The corresponding density of the basis material in the lesion was estimated as the ratio of the measured thickness to the total thickness of the lesion. The error bars represent the standard deviations. This again shows good correlation between the known and measured densities of plastic water.

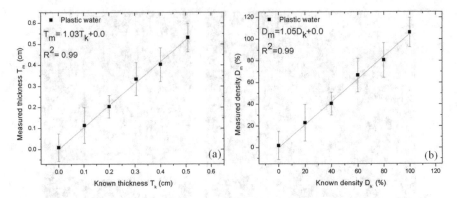

Fig. 10 Measured plastic water thicknesses (**a**) and densities (**b**) as a function of the known values in lesion phantoms embedded in postmortem breasts. The error bars were derived from the standard deviations of the measurements from different breasts

4 Discussions

This study investigates the challenges of characterizing breast lesions by quantifying water and lipid contents to improve the specificity of breast cancer detection during screening mammography. Currently, the quantitative compositional information is not available for suspicious lesions detected in screening mammography. Radiologists mostly depend on the shape, texture, and qualitative signal intensity for determining whether a lesion is benign or malignant. If the chemical composition of a lesion can be provided with high accuracy, it may improve the differentiating power for malignancy. There are only a few studies that have reported the chemical composition of malignant and benign lesions [16, 17, 34]. It is generally understood that the attenuation of malignant lesions is significantly different from that of fibroglandular breast tissue at mammographic energy range due to the difference in their chemical compositions [15, 16, 35]. However, the chemical composition of benign lesions may spread in a relatively large range [34]. While some benign lesions, such as fibroadenomas, may contain relatively high amounts of water, many other types of benign lesions have less water and more lipid content compared to malignant tumors. Although the chemical composition of a lesion may not be a biomarker that can differentiate all malignant tumors from the benign ones, it could help to exclude certain suspicious findings if the decomposed water content is much less than what would be expected for a malignant tumor. By characterizing benign and malignant tumors in terms of their water and lipid contents, this information can be combined with the existing clinical signatures of malignant lesions to improve the accuracy of decision-making. In addition, the proposed technique does not require additional exposures. The compositional information will be available using a normal screening scan. The quantitative and automatic nature of dual-energy decomposition also makes it ready to implement with any computer-aided diagnostic (CAD) program. Therefore, the proposed technique offers a useful addition to current clinical practice and may increase cancer detection rates and reduce the number of false-positives, potentially decreasing the number of negative breast biopsies without additional radiation dose.

Dual-energy imaging can provide accurate quantitative thickness measurements for the two basis materials without relying on any assumptions about the breast thickness model. Therefore, dual-energy imaging can effectively eliminate errors induced by the shape model and the compression paddle's mechanical precision. However, it also limits the compositional analysis to two basis materials. For this reason, we propose to characterize the lesion composition in terms of water and lipid contents, ignoring the small amount of protein. In a previous postmortem study, we have reported that the amount of protein is highly correlated with water and lipid in breast tissues [27]. Therefore, the presence of protein in the lesion is expected to result in a systematic shift in the measured water and lipid thicknesses, which should not affect the precision of the measurement. One advantage of characterizing breast lesions in terms of water and lipid contents is the possibility of performing an accurate calibration, since both basis materials are easy to obtain and have minimal variations. In theory, the system calibration can be done with pure water and oil. However, both of the two basis materials are in a liquid form, which makes it challenging for eventual clinical implementations. In this study, a solid plastic calibration phantom, consisting of plastic water and adipose equivalent materials, was used for the purpose of water-lipid decomposition. For future clinical applications, the decomposed basis material thicknesses can be converted into water and lipid values using a basis transfer function [17]. The plastic phantoms' exact composition, in terms of water and lipid, can be experimentally determined through an initial calibration process by using liquid water and lipid. The measured log signal from the plastic phantoms can be mapped to the log signals obtained for water and lipid of the same thickness. The phantom composition can then be determined using the mapping coefficients. Although such a process still involves the use of liquid, it will have to be done only one time. Once the plastic phantoms' water-lipid composition is determined, the dual-energy calibration can be carried out with plastic-based phantoms by converting the phantom thicknesses into actual water and lipid thicknesses. This is a linear transformation process, which will not affect the precision of the lesions' compositional measurements.

In mammography, the measured signal is determined by the total breast thickness that the photons penetrate before getting registered in the detector, which includes both signal from the lesion and the background tissue. To be able to isolate the information from the lesion, we propose to subtract the signal in the lesion region from the surrounding tissue. The assumption here is that the chemical composition of the normal breast tissue above and below the lesion region is the same as that in the adjacent region. Such assumptions can be easily satisfied in a phantom study with uniform background. In clinical images, the presence of anatomical noise imposes variations of water content over the whole breast. However, the anatomical noise is composed of mostly low spatial frequency components, which means that the change of water thickness is slow and smooth. Therefore, it can be assumed that the background water content around a lesion is approximately uniform, as long as the size of the lesion is small. In this study, a BR3D phantom with swirl pattern was used to simulate the anatomical background. An initial evaluation was conducted to determine the optimal size of the background ROI. The results suggested that an annular ROI with a thickness of 1 mm resulted in the smallest error in plastic water thickness quantification in lesion phantoms. This optimized size was then used in all phantom and tissue studies. A similar optimization process may be conducted for the clinical implementation of the proposed technique.

When comparing the water densities in lesions made from bovine tissue, we observed a good linear correlation between spectral mammography and chemical analysis. However, the RMS error was approximately 7.4%. This is expected, since the proposed method decomposes the measured lesion signal into two basis materials, ignoring the presence of protein. On the other hand, in chemical analysis, tissue samples were characterized in terms of water, lipid, and protein contents. In dual-energy decomposition, the signal from protein will be partly decomposed into water content, leading to higher water density estimation from spectral mammography. Since water and protein concentrations are highly correlated, postmortem breasts with higher water content will get higher signal from protein, which explains the fact that the correlation slope was larger than unity. Nevertheless, as the results suggested, this will not affect the correlation between the measured and the known values, which means that the precision for lesion characterization will be preserved. It should also be noted that although chemical analysis offers a good measure of breast composition, experimental errors may still be present for small breast samples, especially when the amount of water is low. We observed more variance for low-density lesions, which can be partly attributed to the increased experimental error in chemical analysis. Nevertheless, we observed a good linear correlation in water density measurements between spectral mammography and chemical analysis with R^2 comparable to that from the disk phantoms (Fig. 8). More importantly, the SEE was approximately the same as the measurements based on plastic disk lesion phantoms. Thus, it is suggested that the precision of the proposed method is preserved when using only two basis materials for lesion characterization.

The study investigated the effects of breast shape and density on the accuracy of lesion characterization in terms of water and lipid. The results suggested that dual-energy decomposition measurements do not depend on breast thickness, glandularity, or lesion depth. The mean values in water density measurements were in good agreement with the known values within the range of RMS errors shown in Table 1. The variation induced by breast characteristics was also comparable to the measurement precision. This suggested that the accuracy and precision will not be affected by breast shape and density.

The current study used a clinical imaging protocol that was designed for screening mammography. The high-energy threshold was already optimized for breast density quantification according to a previous simulation study [25] and is expected to offer a good SNR for lesion characterization in terms of water and lipid contents. The mean glandular dose was estimated to be 0.2–0.8 mGy using automatic exposure control, which is lower than the current clinical standard [36]. The method aims to offer a quantitative metric to characterize the chemical composition of a suspicious lesion in screening mammography without additional radiation dose. At the current dose level, the noise can be amplified through the dual-energy decomposition process. The decomposition noise will increase the standard deviation of the measurement in an ROI, but may not significantly affect the mean values in the measurement, as long as a sufficient number of pixels can be included in the ROI. Thus, a reliable characterization of lesion composition may be obtained at a relatively low dose level. However, for very small lesions, the accuracy of lesion characterization may be impaired by the small difference between lesion and normal tissue in the background. A limit of 10 mm was predicted by a previous theoretical study on lesion discrimination [37]. In clinical images, lesions may have irregular shapes. This method measures the integrated signal inside a lesion, followed by the subtraction of the background. Thus, it is less sensitive to the shape but more sensitive to the integrated volume of a lesion.

5　Conclusions

In summary, dual-energy material decomposition method was used to characterize lesions in terms of water and lipid in a phantom study with a photon-counting spectral mammography system. The proposed method can accurately quantify the water content with an approximate error of less than 0.4 mm. The results showed that the technique is accurate for different breast thicknesses, densities, and lesion locations. Accurate quantification of breast lesion content can potentially help distinguish between malignant and benign lesions.

Acknowledgments The authors would like to thank Dr. Hyo-Min Cho, David Sennung, Nikita Kumar, and Drs. Bahman Sadeghi, Hanna Javan and Alfonso Lam Ng for their support in experiment and data analysis. The authors would like to thank Dr. Erik Fredenberg for valuable technical support and acknowledge Philips Medical Systems for providing the MicroDose mammography system for this research. This work was supported in part by NIH/NCI grant R01CA13687.

References

1. Dean P, Pamilo M. Large-scale breast cancer screening with mammography: high cancer detection rates combined with a low risk of benign biopsy. Radiology. 1997;Suppl 205:143.
2. Schmitt E, Threatt B. Effective breast cancer detection with film-screen mammography. J Can Assoc Radiol. 1985;36:304–7.
3. D'Orsi CJ. Early detection of breast cancer: mammography. Breast Cancer Res Treat. 1991;18:S107–9.
4. Lewin JM, Hendrick RE, D'Orsi CJ, Isaacs PK, Moss LJ, Karellas A, Sisney GA, Kuni CC, Cutter GR. Comparison of full-field digital mammography with screen-film mammography for cancer detection: results of 4,945 paired examinations 1. Radiology. 2001;218:873–80.
5. Sala M, Comas M, Macià F, Martinez J, Casamitjana M, Castells X. Implementation of digital mammography in a population-based breast cancer screening program: effect of screening round on recall rate and cancer detection. Radiology. 2009;252:31–9.

6. Bird RE, Wallace TW, Yankaskas BC. Analysis of cancers missed at screening mammography. Radiology. 1992;184:613–7.
7. Jiang Y, Nishikawa RM, Schmidt RA, Metz CE, Giger ML, Doi K. Improving breast cancer diagnosis with computer-aided diagnosis. Acad Radiol. 1999;6:22–33.
8. Wu Y, Giger M, Doi K, Vyborny C, Schmidt B, Metz C. Artificial neural networks in mammography: application to decision making in the diagnosis of breast cancer. Radiology. 1993;187:81–7.
9. Woods RW, Sisney GS, Salkowski LR, Shinki K, Lin Y, Burnside ES. The mammographic density of a mass is a significant predictor of breast cancer. Radiology. 2011;258:417.
10. Chung S, Cerussi A, Klifa C, Baek H, Birgul O, Gulsen G, Merritt S, Hsiang D, Tromberg B. In vivo water state measurements in breast cancer using broadband diffuse optical spectroscopy. Phys Med Biol. 2008;53:6713.
11. Hsiang D, Durkin A, Butler J, Tromberg BJ, Cerussi A, Shah N. In vivo absorption, scattering, and physiologic properties of 58 malignant breast tumors determined by broadband diffuse optical spectroscopy. J Biomed Opt. 2006;11:044005.
12. Tromberg BJ, Cerussi A, Shah N, Compton M, Durkin A, Hsiang D, Butler J, Mehta R. Imaging in breast cancer: diffuse optics in breast cancer: detecting tumors in pre-menopausal women and monitoring neoadjuvant chemotherapy. Breast Cancer Res. 2005;7:279.
13. Haka AS, Shafer-Peltier KE, Fitzmaurice M, Crowe J, Dasari RR, Feld MS. Diagnosing breast cancer by using Raman spectroscopy. Proc Natl Acad Sci U S A. 2005;102:12371–6.
14. Olmstead EG. Mammalian cell water; physiologic and clinical aspects. Philadelphia: Lea & Febiger; 1966.
15. Johns PC, Yaffe MJ. X-ray characterization of normal and neoplastic breast tissues. Phys Med Biol. 1987;32:675–95.
16. Tomal A, Mazarro I, Kakuno EM, Poletti ME. Experimental determination of linear attenuation coefficient of normal, benign and malignant breast tissues. Radiat Meas. 2010;45:1055–9.
17. Fredenberg E, Dance DR, Willsher P, Moa E, von Tiedemann M, Young KC, Wallis MG. Measurement of breast-tissue x-ray attenuation by spectral mammography: first results on cyst fluid. Phys Med Biol. 2013;58:8609–20.
18. Ducote JL, Molloi S. Quantification of breast density with dual energy mammography: a simulation study. Med Phys. 2008;35:5411–8.
19. Ducote JL, Molloi S. Quantification of breast density with dual energy mammography: an experimental feasibility study. Med Phys. 2010;37:793–801.
20. Molloi S, Ducote JL, Ding H, Feig SA. Postmortem validation of breast density using dual-energy mammography. Med Phys. 2014;41:081917.
21. Lundqvist M, Danielsson M, Cederstroem B, Chmill V, Chuntonov A, Aslund M. Measurements on a full-field digital mammography system with a photon counting crystalline silicon detector. Proc SPIE Int Soc Opt Eng. 2003;5030:547–52.
22. Yadava GK, Kuhls-Gilcrist AT, Rudin S, Patel VK, Hoffmann KR, Bednarek DR. A practical exposure-equivalent metric for instrumentation noise in x-ray imaging systems. Phys Med Biol. 2008;53:5107–21.
23. Cho H-M, Barber WC, Ding H, Iwanczyk JS, Molloi S. Characteristic performance evaluation of a photon counting Si strip detector for low dose spectral breast CT imaging. Med Phys. 2014;41:091903.
24. Wang AS, Harrison D, Lobastov V, Tkaczyk JE. Pulse pileup statistics for energy discriminating photon counting x-ray detectors. Med Phys. 2011;38:4265–75.
25. Ding HJ, Molloi S. Quantification of breast density with spectral mammography based on a scanned multi-slit photon-counting detector: a feasibility study. Phys Med Biol. 2012;57:4719–38.
26. Laidevant AD, Malkov S, Flowers CI, Kerlikowske K, Shepherd JA. Compositional breast imaging using a dual-energy mammography protocol. Med Phys. 2009;37:164–74.
27. Johnson T, Ding H, Molloi S. Breast density quantification with breast computed tomography (CT): a post-mortem study. Phys Med Biol. 2013;58:8573–91.

28. Cho HM, Ding H, Kumar N, Sennung D, Molloi S. Calibration phantoms for accurate water and lipid density quantification using dual energy mammography. Phys Med Biol. 2017;62:4589–603.
29. Fredenberg E, Lundqvist M, Cederström B, Åslund M, Danielsson M. Energy resolution of a photon-counting silicon strip detector. Nucl Instrum Methods Phys Res A. 2010;613:156–62.
30. Åslund M, Cederström B, Lundqvist M, Danielsson M. Scatter rejection in multislit digital mammography. Med Phys. 2006;33:933–40.
31. Ducote JL, Klopfer MJ, Molloi S. Volumetric lean percentage measurement using dual energy mammography. Med Phys. 2011;38:4498–504.
32. Cardinal HN, Fenster A. An accurate method for direct dual-energy calibration and decomposition. Med Phys. 1990;17:327–41.
33. Levenberg K. A method for the solution of certain problems in least squares. Q Appl Math. 1944;2:164–8.
34. Sha LW, Ward ER, Stroy B. A review of dielectric properties of normal and malignant breast tissue. In: Proceedings IEEE SoutheastCon 2002. IEEE; 2002. p. 457–62.
35. Carroll FE, Waters JW, Andrews WW, Price RR, Pickens DR, Willcott R, Tompkins P, Roos C, Page D, Reed G, Ueda A, Bain R, Wang P, Bassinger M. Attenuation of monochromatic X-rays by normal and abnormal breast tissues. Investig Radiol. 1994;29:266–72.
36. Hendrick RE, Pisano ED, Averbukh A, Moran C, Berns EA, Yaffe MJ, Herman B, Acharyya S, Gatsonis C. Comparison of acquisition parameters and breast dose in digital mammography and screen-film mammography in the American College of Radiology Imaging Network digital mammographic imaging screening trial. Am J Roentgenol. 2010;194:362–9.
37. Norell B, Fredenberg E, Leifland K, Lundqvist M, Cederstrom B. Lesion characterization using spectral mammography. In: Medical imaging 2012: physics of medical imaging, vol. 8313. SPIE; 2012.

Quantitative Breast Imaging with Low-Dose Spectral Mammography

Huanjun Ding and Sabee Molloi

1 Introduction

Breast cancer is the most common cancer in women and the second leading cause of cancer death [1]. It is well known that an important risk factor for breast cancer is mammographic breast density. Breast carcinoma develops in the glandular component of the breast, and previous reports have shown that women who have mammographic density of 75–100% have a four- to fivefold increase in breast cancer risk as compared to those with mammographic density of 0–25% [2–4]. Furthermore, the sensitivity of screening mammography decreases with increases in mammographic density [3, 5–12]. Therefore, it is important to accurately measure breast density in order to estimate the overall risk of breast cancer. The current clinical standard for breast density assessment involves visual assessment of mammograms, which is known to be limited by inter- and intra-observer limitations [13–15]. There have been a number of previous reports of more quantitative approaches for measuring breast density [16–18]. However, the fundamental limitation of these techniques is estimation of glandular fraction from a single-energy mammogram. These techniques often require assumptions such as breast thickness for breast density estimation [19–22]. There is currently no available reference standard for validating the accuracy of breast density in patients. The currently available breast density techniques have not been validated for accuracy using a reference gold standard such us chemical analysis in postmortem breasts. Therefore, the accuracy of these techniques is not known.

H. Ding · S. Molloi (✉)
Department of Radiological Sciences, University of California, Irvine, CA, USA
e-mail: symolloi@uci.edu

© The Author(s), under exclusive license to Springer Nature Switzerland AG 2023
S. Hsieh, K. (Kris) Iniewski (eds.), *Photon Counting Computed Tomography*,
https://doi.org/10.1007/978-3-031-26062-9_6

It is possible to accurately measure breast density using dual-energy mammography without any assumptions for breast thickness since low- and high-energy images are used to measure glandular and adipose thicknesses.

Dual-energy decomposition requires a prior system calibration step to generate the necessary basis material-specific decomposition coefficients [23]. Previous studies have used the available glandular and adipose plastic phantoms for dual-energy calibration [24–31]. However, these phantoms were not designed for dual-energy mammography since the chemical composition of glandular and adipose tissues is not known with the required accuracy for dual-energy material decomposition. Although measured breast density was accurate in phantom studies [30], the breast density measurements in bovine tissue and postmortem breasts resulted in a large systematic error [24, 29]. Liquid phantoms have previously been investigated to address this limitation [32]. However, liquid phantoms are not practical for routine clinical application. To address this limitation, dual-energy calibration can be performed with plastic water and adipose-equivalent plastic phantom and then employ a conversion process to generate the true water and lipid thicknesses.

In this chapter, we have first discussed a new dual-energy calibration technique for spectral mammography. We have evaluated the improvement of the accuracy in breast density quantification using water and lipid as basis materials. The feasibility of quantifying breast contents using conversion coefficient from plastic water and adipose phantoms into water and lipid dual-energy basis materials was investigated. Validation studies were performed with water lipid mixture phantom and postmortem breasts. Next, we have applied this technique to validate the accuracy of water and lipid density quantification in postmortem breasts with dual-energy mammography using chemical analysis as the reference standard.

2 Theory

2.1 Theoretical Conversion

In general, breast tissue of a given thickness ($\mu_B t_B$) can be decomposed into any two basis materials of different and low atomic numbers. In this study, we have used two sets of basis materials: (1) plastic water- (μ_{pw}) and adipose tissue-equivalent material (μ_{ap}) and (2) water (μ_w) and lipid (μ_l). In a matrix notation, this can be written as

$$\mu_B t_B = \mathbf{T_p} \boldsymbol{\mu_p} = \mathbf{T_{wl}} \boldsymbol{\mu_{wl}}, \quad \text{where} \tag{1}$$

$$\mathbf{T_p} = \begin{bmatrix} t_{pw} & t_{ap} \end{bmatrix}, \boldsymbol{\mu_p} = \begin{bmatrix} \mu_{pw} \\ \mu_{ap} \end{bmatrix}, \mathbf{T_{wl}} = \begin{bmatrix} t_w & t_l \end{bmatrix}, \boldsymbol{\mu_{wl}} = \begin{bmatrix} \mu_w \\ \mu_l \end{bmatrix}. \tag{2}$$

Matrices μ_p and μ_{wl} are related through a conversion matrix A_{wl}, which contains the equivalent thicknesses of water and lipid for the CIRS tissue-equivalent calibration phantoms:

$$A_{wl} = \begin{bmatrix} a_{pw}^w & a_{pw}^l \\ a_{ap}^w & a_{ap}^l \end{bmatrix}. \tag{3}$$

In an ideal case, A_{wl} can be determined through simulations using the theoretical attenuation coefficients of the basis materials. With this notation, the relationship between the true water and lipid contents in breast and the measured data from phantom calibration becomes

$$T_p\mu_p = T_{wl}\mu_{wl} = T_{wl}A_{wl}\mu_p \Rightarrow T_{wl} = T_pA_{wl}^{-1}. \tag{4}$$

The effective linear attenuation coefficients (μ_{eff}) of the base materials were calculated for low- and high-energy range based on simulated polyenergetic x-ray spectra before and after passing through certain thickness of tissue using Eqs. (5) and (6), respectively [33]:

$$\mu_{eff-low} = \ln\left(\int_{E_{min}}^{E_s} I_0\ (E)\cdot \eta(E)dE \middle/ \int_{E_{min}}^{E_s} I(E)dE\right)/T \tag{5}$$

$$\mu_{eff-high} = \ln\left(\int_{E_s}^{E_{max}} I_0\ (E)\cdot \eta(E)dE \middle/ \int_{E_s}^{E_{max}} I(E)dE\right)/T \tag{6}$$

where $I_0(E)$ is the incident x-ray spectrum which represents the emitted number of photons per unit energy at energy E, $\eta(E)$ is the energy response function of the detector, and $I(E)$ is the detected x-ray spectrum, which was simulated using the TASMIP code [34]. Energy spectra were simulated from 7 keV (E_{min}) which is the noise floor for the investigated detector up to the maximum photon energy (E_{max}) for 26, 29, 32, 35, and 38 kVp beams. The calculated linear attenuation coefficient curves based on the chemical composition of the base materials are presented in Fig. 1. The tube voltage is automatically determined by the system based on the breast thickness (T). The high-energy threshold (E_s) for dual energy was automatically selected based on a built-in calibration of the spectral mammography system. The calculated effective linear attenuation coefficients of each material for five different tube voltages are shown in Table 1. The theoretical linear conversion coefficients were calculated based on Table 1.

Fig. 1 The linear attenuation coefficient of pure water, lipid, plastic water, and CIRS adipose-equivalent material

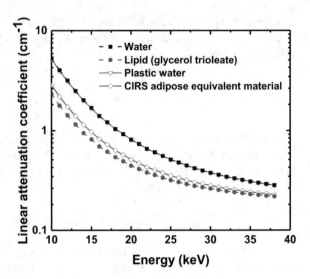

Table 1 The effective linear attenuation coefficient of four materials at low and high energy of five different kVp

	Water		Lipid		Plastic water		Adipose equivalent	
kVp	Low	High	Low	High	Low	High	Low	High
26	1.351	0.708	0.719	0.405	1.346	0.708	0.840	0.461
29	1.160	0.595	0.636	0.360	1.158	0.595	0.736	0.403
32	1.011	0.515	0.570	0.327	1.011	0.515	0.655	0.361
35	0.901	0.454	0.520	0.301	0.902	0.454	0.593	0.329
38	0.762	0.398	0.459	0.275	0.762	0.399	0.517	0.297

2.2 Experimental Conversion

Nonlinear rational fitting function was used in this study to calculate calibration parameters $(a_0, a_1, \dots)$ by substituting the known material thickness values, t_i, and the corresponding dual-energy log-signal S_L and S_H of tissue-equivalent plastic calibration phantom in Eq. (7):

$$ t_i = \frac{a_0 + a_1 S_L + a_2 S_H + a_3 S_L^2 + a_4 S_L S_H + a_5 S_H^2 + a_6 S_L^3 + a_7 S_H^3}{1 + b_1 S_L + b_2 S_H + b_3 S_L^2 + b_4 S_H^2} \tag{7} $$

The composition of an object in terms of plastic water and adipose tissue-equivalent material thickness was estimated by using the calibration parameters.

For experimental conversion coefficient, the log-signal of empty cylindrical plastic container was decomposed into plastic water and adipose tissue-equivalent material. These thicknesses were subtracted from the estimated thickness to calculate the thickness of plastic water and adipose tissue-equivalent phantom of pure water and lipid.

The water (T_w) and lipid (T_l) thickness can be represented as follows:

$$T_i = a_0 + a_1 t_{ap} + a_2 t_{pw} \tag{8}$$

During the calibration process using pure water and lipid, the known thickness (T_w, T_l) and calculated thickness of plastic water (t_{pw}) and adipose tissue-equivalent phantom (t_{ap}) from dual-energy calibration process were substituted. Afterward, the linear system of equations was solved to determine conversion coefficients of a_0, a_1, and a_2. In the validation process, unknown water and lipid thicknesses in the mixture phantom and postmortem were estimated by applying the conversion coefficient to the calculated plastic water and adipose tissue-equivalent material thickness.

3 Dual-Energy Material Decomposition Using Spectral Mammography: A Simulation Study

A previously reported analytical simulation model was used in this study [27, 31, 35]. In summary, the simulation model considers photon emission from the x-ray source, attenuation through the object, and subsequent absorption in the detector. The attenuation coefficients of each material in the low- and high-energy bin were calculated, and the detector signals and their uncertainties due to statistical x-ray noise were recorded. Polyenergetic x-ray spectra were provided by the TASMIP code [34]. The experimental setup was simulated with a 0.75 mm of aluminum pre-filtration. The simulation was performed to investigate the uncertainty of measurement according to the basis decomposition materials. The attenuation coefficient of breast tissue with low and high energy was calculated based on previous data for postmortem breast tissue composition ratio of water, lipid, and protein [36]. Various thicknesses were simulated from 1 to 8 cm in intervals of 1 cm, which cover the clinically compressed breast tissue range. The quantum detection efficiency (QDE) of the Si-based photon-counting detector was simulated by considering 3.6-mm-thick Si crystal's x-ray attenuation property, which is the effective thickness of the Si-strip detector for the spectral mammography system in the experimental studies. The tube voltage and the high-energy threshold were selected based on a built-in lookup table from the spectral mammography system. Nonlinear rational fitting function was used for dual-energy calibration. The simulation tested the performance of two sets of basis materials: in the first set, pure water and lipid were used as the decomposition basis; in the second set, tissue-equivalent plastic water and adipose phantoms were chosen to be the decomposition basis. As shown in Fig. 1, the attenuation property of plastic water is in good agreement with water, but adipose-equivalent material is not in good agreement with lipid since in addition to lipid, adipose tissue contains certain amount of water.

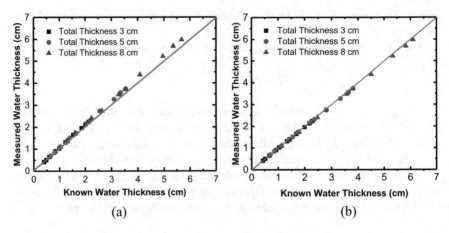

Fig. 2 The simulation results of water thickness measurements using (**a**) pure water and lipid as basis materials for calibration and (**b**) compensate protein contents by adding 58% of the known protein contents into water. The solid lines represent line of identity in both figures

In the simulation study, breast tissue, consisting of water, lipid, and protein with various volumetric percentages derived from a previous postmortem study, was decomposed into water and lipid basis materials. In Fig. 2a, the decomposed water thicknesses were slightly overestimated in comparison to the known values. This is due to the fact that a certain percentage of protein has been assigned into water. Nevertheless, there is a good linear correlation with a slope of 1.05 and an offset of 0.01 cm. This systematic error can be adjusted by removing the contribution of protein in the decomposed water thickness. Using the method discussed in section II.F, we have estimated that protein can be decomposed into 58% of water and 42% of lipid. Figure 2b shows the correlation between decomposed water thicknesses and the adjusted water thickness after adding 58% of known protein contents. The correlation slope was derived to be 0.98. The averaged root-mean-square (RMS) error was estimated to be 0.06 cm. The difference between known and estimated water thickness tends to increase according to the thickness of the breast, but it wasn't enough to show an obvious difference. In the second simulation study, the same breast samples as used in the previous study were decomposed into plastic water and adipose-equivalent phantoms. Figure 3a shows that water thicknesses were underestimated when breast tissues were decomposed into plastic water and adipose tissue-equivalent material. This can be explained by the fact that the adipose tissue-equivalent material includes a certain percent of water [32]. To test this assumption, the theoretically and experimentally acquired conversion coefficients, used to convert the plastic water and adipose tissue-equivalent material thickness into the water and lipid thickness, were used for the simulated results as shown in Fig. 3b, c. The correlation was significantly improved after the conversion. For the theoretical conversion, the slope of the linear fitting was 0.95. The offsets for the linear fittings were 0.29, 0.48, and 0.77 cm for breast thicknesses of 3, 5, and

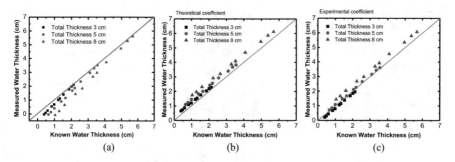

Fig. 3 The simulation results of (**a**) water thickness measurements using plastic water and adipose tissue-equivalent materials as basis materials for calibration. The converted water thickness by applying (**b**) theoretically and (**c**) experimentally acquired conversion coefficients. The solid lines represent line of identity in all figures

8 cm, respectively. For the experimental conversion, the slope of the linear fitting was 1.00. The offsets of the linear fittings were -0.17, 0.06, and 0.40 cm for breast thicknesses of 3, 5, and 8 cm, respectively. The RMS errors were derived to be 0.4 cm and 0.22 cm for the theoretical and experimental conversions, respectively. The simulation results show that the error can be reduced when water and lipid are used as basis materials for dual-energy calibration.

4 Dual-Energy Breast Density Quantification Using Spectral Mammography: A Phantom Study

4.1 Spectral Mammography System

All the experimental data was acquired using a spectral mammography system (MicroDose SI L50, Phillips Inc.). The system is composed of a tungsten-target x-ray tube, aluminum filter, a pre- and a post-collimator, and Si-strip detector units, all mounted on a common arm (see Fig. 4). Si-strip detectors are placed in an edge-on geometry as an image receptor to improve quantum efficiency of silicon. Application-specific integrated circuits (ASIC) are used in which photon energies below a low-energy threshold are considered as noise, and all remaining photons are sorted into two energy bins using a high-energy threshold. These characteristics facilitate dual-energy image acquisition without electronic readout noise within a single exposure. The pre- and post-collimator effectively reject scattered radiation to less than 6%, which can improve accuracy of the material decomposition technique [37]. A more detailed explanation regarding the spectral mammography system characteristics can be found in previous reports [38, 39].

Fig. 4 Dual-energy mammography system based on a Si-strip photon-counting detector in a scanned multi-slit geometry. The system consists of a tungsten-anode x-ray tube, an aluminum filter, a pre- and a post-collimator, and the detector unit

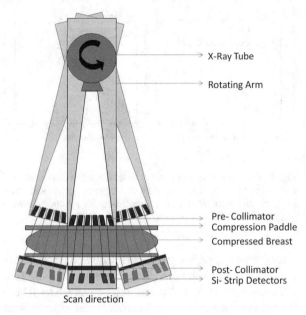

4.2 Tissue-Equivalent Phantoms

The tissue-equivalent plastic phantom for dual-energy calibration consists of water and adipose tissue-equivalent material. Breast tissue can be decomposed into water, lipid, and protein with average percentage volumes of 34.2%, 59.9%, and 5.8%, respectively [36, 40]. However, dual-energy mammography can be used to decompose two different materials so the calibration phantom was designed to include only water and lipid. Plastic water (CIRS, Norfolk, VA) was formulated to mimic x-ray attenuation properties of water within 1% error for the mammographic energy range. However, it is not possible to manufacture a solid plastic phantom simulating lipid. It is possible to use an existing adipose tissue-equivalent phantom (CIRS, Norfolk, VA) which is equivalent to approximately 85% lipid and 15% water. Figure 5 shows a schematic diagram of the tissue-equivalent calibration phantom. Combinations of the two materials were used to produce four different total thicknesses: 2, 4, 6, and 8 cm. Each thickness includes five different areas whose plastic water densities range from 0% to 100% with an increasing step of 25%. The phantom thickness and water densities were selected to include clinically relevant breast thickness and densities by providing 20 calibration points. Five different tube voltages of 26, 29, 32, 35, and 38 kVp were used to acquire log-signal from the simultaneously acquired low- and high-energy images of the phantom. The low- and high-energy log-signals of an object can be expressed as a linear combination of the plastic water and adipose tissue-equivalent material [31, 41].

We have also acquired phantom images using pure water and lipid in their liquid form. The low- and high-energy images of liquid water and lipid were measured in

Fig. 5 Schematic diagram of the calibration phantom. The phantom includes 20 combinations of plastic water and adipose tissue-equivalent phantom

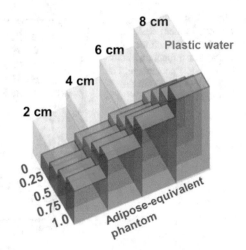

seven different thicknesses from 0 to 6 cm, respectively. The liquid was poured into a cylindrical plastic container. The cross-sectional area of the container was carefully measured with a caliper and averaged over three independent measurements. The thickness of liquid was then calculated using the measured weight differences in between each exposure, the reported room temperature density, and the averaged cross-sectional area. The low- and high-energy images were acquired at 26, 29, 32, 35, and 38 kVp. The log-signal of each material in different thicknesses was used to calculate the conversion coefficient from plastic water and adipose tissue-equivalent material to water and lipid using linear fitting. The evaluation of the conversion coefficient was performed by using a mixture of two materials in different combinations.

4.3 Results

Figure 6 shows the converted water percentage using theoretically calculated conversion coefficients as a function of the known water percentage in the validation process. The result shows high RMS errors as shown in Table 2. Figure 7 shows the experimentally calibrated water percentage at five different tube voltages as a function of the known water percentage. There is an excellent agreement between the calibrated and known water percentages with substantially reduced RMS errors as shown in Table 2.

Dual-energy mammography can be used to decompose only two different materials. Aluminum and polymethyl methacrylate (PMMA) [41–45] have previously been used as the basis materials for dual-energy decomposition. However, the required aluminum thickness for dual-energy mammography calibration is relatively small, which can introduce calibration error. Furthermore, the measured aluminum and PMMA thicknesses have to be converted to materials relevant to

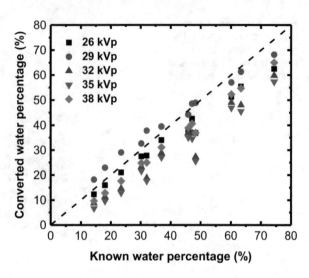

Fig. 6 The theoretically converted water percentage with five different tube voltages as a function of the known water percentage. The dotted lines represents the line of identity

Table 2 RMS errors between the converted and known water densities in mixture phantom for validation with two different methods

RMS error	26 kVp	29 kVp	32 kVp	35 kVp	38 kVp	Average
Theoretical	6.7%	3.8%	11.7%	13.6%	7.3%	8.6%
Experimental	3.0%	1.6%	1.1%	1.1%	1.1%	1.6%

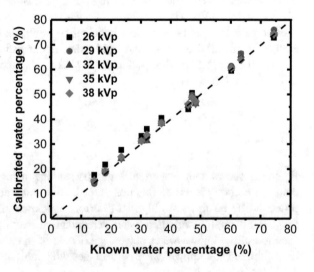

Fig. 7 The experimentally calibrated water percentage at five different tube voltages as a function of the known water percentage. The dotted line represents the line of identity

tissue components. Our results show that water and lipid could be used for dual-energy material decomposition. There is an existing water-equivalent phantom for calibration purposes. However, it is not feasible to manufacture a lipid-equivalent phantom. Therefore, an adipose-equivalent phantom was used for calibration purposes. The adipose phantom is composed of lipid with a small amount of water that was experimentally determined using liquid water and lipid as the calibration materials.

The advantage of using water and lipid as calibration materials is that they can be directly related to the fundamental components of breast tissue. Furthermore, water density may also be a better representation for lesion characterization. The disadvantage of using water and lipid as the calibration materials is that the protein content of tissue will introduce some systematic error in the measured water and lipid thicknesses. However, protein is a small component of breast tissue. Our results show that the systematic error due to the protein content in tissue is relatively small.

The phantom study suggested that experimentally acquired conversion coefficients can provide accurate measurement of breast tissue composition. Conversion of the dual-energy measurements into water and lipid thicknesses improves the accuracy in tissue decomposition studies.

5 Quantification of Water and Lipid Density with Dual-Energy Mammography: Validation in Postmortem Breasts

5.1 Image Acquisition and Processing

Twenty pairs (left and right) of postmortem breast were acquired from the Willed Body Program in University of California, Irvine. The postmortem breasts were surgically removed to the pectoralis major muscle and were kept in plastic wrap for the entire imaging process to minimize water loss. The range of breast mass, compressed thickness, and measured volumetric breast density were 154–2368 g, 1.2–9.0 cm, and 13–70%, respectively. Depending on the compressed thickness of the postmortem breasts, the images were acquired using 26–35 kVp and 10–15 mAs. For postmortem breasts with compressed thickness of up to 3.1 cm, 3.2–4.9 cm, 5.0–6.9 cm, and higher than 7 cm, images were acquired at 26 kVp and 10 mAs, 29 kVp and 10 mAs, 32 kVp and 11 mAs, and 35 kVp and 15 mAs, respectively.

Dual-energy material decomposition of the low- and high-energy images was used to produce individual pixel measurements of water and lipid material thickness. The decomposition was based on calibration with water-equivalent and adipose-equivalent phantoms. Dual-energy mammography was used to perform material decomposition using two basis materials. However, tissue is composed of three main components of water, lipid, and protein. Since the protein content will be decomposed into water and lipid bases, the expected water density was also adjusted by adding 58% of the protein content to the measured water density from chemical analysis. The details of calculating the water and lipid density have been previously reported [28].

We have previously compared a number of different techniques for breast density measurement [24, 46]. In this study, we also tried to use a commercially available

breast density technique (Quantra, Hologic Inc., Bedford, MA) for comparison purposes. However, existing commercial breast density techniques (Quantra, Hologic
Inc., Bedford, MA and Volpara Solutions, Wellington, New Zealand) require a
shape model to estimate breast thickness in the periphery of the breast where the
compression paddle is not in contact with the breast [47, 48]. However, it is not
possible to measure breast density in postmortem breasts using these techniques
since the existing shape models are not applicable for postmortem breasts.

5.2 Chemical Analysis

After completion of image acquisition, all the postmortem breast specimens were
chemically decomposed into their water, lipid, and protein contents. The postmortem breasts were weighed before and after the completion of image acquisition
and any mass change during imaging was attributed to water loss. This amount
was added to the final water fraction. The breast specimens were cut into pieces
of approximately $5 \times 5 \times 5$ mm^3 and placed into a vacuum oven at 95 °C for
approximately 48 hours. Water content was calculated as the reduction in tissue
mass. To separate the remaining lipid and protein components, the dried tissue
was mixed with petroleum ether, grounded and agitated at room temperature for
approximately 1 hour, to dissolve the lipid content. The mixture was then vacuum
filtered with a Buchner funnel, while petroleum ether was poured over the mixture
to filter out any residual lipid content, with the assumption that the petroleum
ether solution contained the entire breast lipid. The lipid weight was determined
by evaporating the petroleum ether. Remaining protein component in the filter
was placed back in the vacuum oven for 30 minutes to evaporate the petroleum
ether. After evaporation, the remaining materials were protein and small amount of
minerals, such as Ca, which were found to be insignificant as compared to the total
weight of the specimens [40]. Water, lipid, and protein masses were converted into
volumes using their known densities of 1, 0.924, and 1.35 g/cm^3, respectively.

5.3 Results

Examples of standard mammography image along with water and lipid dual-energy
decomposed images are shown in Fig. 8. The linear correlation of water (W)
and lipid (L) density from two different views, simulating craniocaudal (CC) and
mediolateral oblique (MLO) views, using dual-energy mammography are shown in
Fig. 9a, c, respectively. The measurements for water density from view 1 (W_{V1}) and
view 2 (W_{V2}) were related by $W_{V2} = 1.02\ W_{V1} - 0.55\%$ ($r = 1.00$). In addition,
in a Bland–Altman plot, the mean difference between water density from the two
views was $0 \pm 2.5\%$ (Fig. 9b). The measurements for lipid density from view 1
(L_{V1}) and view 2 (L_{V2}) were related by $L_{V2} = 1.02\ L_{V1} - 1.08\%$ ($r = 1.00$). In

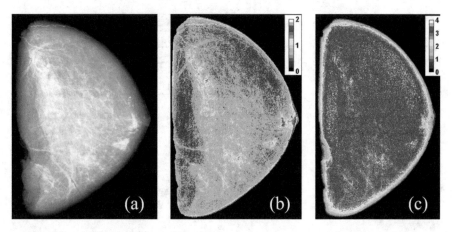

Fig. 8 Examples of standard mammography image (**a**) along with color maps representing water (**b**) and lipid (**c**) thicknesses (in cm) in a given pixel

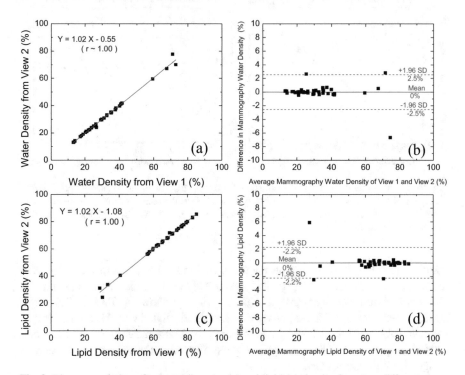

Fig. 9 Linear correlation of volumetric water (**a**) and lipid (**c**) density from two different views using dual-energy mammography measurements. In addition, in Bland–Altman plots, the mean differences between water and lipid density from the two views were $0 \pm 2.5\%$ and $0 \pm 2.2\%$, respectively (**b**, **d**). SD standard deviation

Table 3 Summary of the linear regression analysis of water and lipid density from two different views using dual-energy mammography measurements

	Water	Lipid
Slope	1.02 ± 0.01	1.02 ± 0.01
Intercept	$-0.55 \pm 0.49\%$	$-1.08 \pm 0.91\%$
Pearson's r	1.00	1.00
SEE	1.25%	1.11%

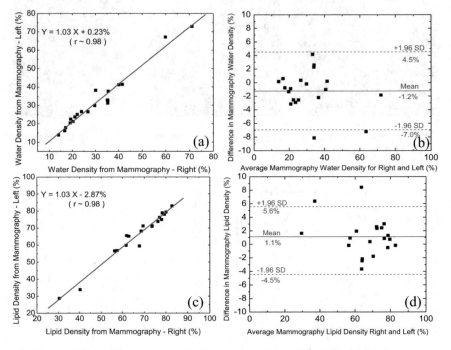

Fig. 10 Linear correlation of water (**a**) and lipid (**c**) density from dual-energy mammography from the right and left breasts. In addition, in Bland–Altman plots, the mean differences between water and lipid density from the right and left breasts were $-1.2 \pm 2.9\%$ and $1.1 \pm 2.8\%$, respectively (**b**, **d**). SD standard deviation

addition, in a Bland–Altman plot, the mean difference between lipid density from the two views was $0 \pm 2.2\%$ (Fig. 9d). A summary of the linear regression analysis of water and lipid density from two different views is shown in Table 3. The water and lipid density from different views were highly correlated with a standard error of estimate (SEE) of 1.25% and 1.11%, respectively. These measurements are a good estimate of precision of dual-energy mammography since the same breast tissue was imaged from different views.

Another estimate of precision for measuring water and lipid density was evaluated by comparing measurements from contralateral breasts. In this case, the biological variance between contralateral breasts is also included in the measurement. The relation of dual-energy water density from the right breast (W_{DE-R}) and dual-energy water density from the left breast (W_{DE-L}) is shown in Fig. 10a. The

Table 4 Summary of the linear regression analysis between right and left water and lipid density from dual-energy measurements

	Water	Lipid
Slope	1.03 ± 0.05	1.03 ± 0.05
Intercept	$0.23 \pm 1.58\%$	$-2.87 \pm 3.35\%$
Pearson's r	0.98	0.98
SEE	2.81%	2.74%

measurements were related by $W_{DE\text{-}L} = 1.03\ W_{DE\text{-}R} + 0.23\%$ ($r = 0.98$). The relation of dual-energy lipid density from the right breast ($L_{DE\text{-}R}$) and dual-energy lipid density from the left breast ($L_{DE\text{-}L}$) is shown in Fig. 10c. The measurements were related by $L_{DE\text{-}L} = 1.03\ L_{DE\text{-}R} - 2.87\%$ ($r = 0.98$). In addition, in Bland–Altman plots, the mean differences between water and lipid density from the right and left breasts were $-1.2 \pm 2.9\%$ and $1.1 \pm 2.8\%$, respectively (Fig. 10b, d). A summary of the linear regression analysis between right and left water and lipid density from dual-energy measurements is shown in Table 4. The water and lipid density from contralateral breasts were highly correlated with a standard error of estimate of 2.81% and 2.74%, respectively. The comparison of water, lipid, and protein from contralateral breasts using chemical analysis provided a separate measure for estimate of precision. The relation of chemical analysis water density from the right breast ($W_{CA\text{-}R}$) and the left breast ($W_{CA\text{-}L}$) is shown in Fig. 11a. The measurements were related by $W_{CA\text{-}L} = 0.96\ W_{CA\text{-}R} + 1.48\%$ ($r = 0.99$). The relation of chemical analysis lipid density from the right breast ($L_{CA\text{-}R}$) and the left breast ($L_{CA\text{-}L}$) is shown in Fig. 11c. The measurements were related by $L_{CA\text{-}L} = 0.98\ L_{CA\text{-}R} + 0.84\%$ ($r = 0.99$). The relation of chemical analysis protein density from the right breast ($P_{CA\text{-}R}$) and the left breast ($P_{CA\text{-}L}$) is shown in Fig. 11e. The measurements were related by $P_{CA\text{-}L} = 0.98\ P_{CA\text{-}R} + 0.30\%$ ($r = 0.98$). In addition, in Bland–Altman plots, the mean differences between water, lipid, and protein density from the right and left breasts were $0.45 \pm 2.0\%$, $0.8 \pm 2.2\%$, and $-0.2 \pm 0.6\%$, respectively (Fig. 11b, d, f). A summary of the linear regression analysis between right and left water, lipid, and protein from chemical analysis measurements is shown in Table 5. The water, lipid, and protein density from right and left breasts were also highly correlated with a standard error of estimate of 1.90%, 2.15%, and 0.54%, respectively. In order to assess the biological variability of the breasts, the masses of the right and left breasts were correlated in Fig. 12. The relation of the right (M_R) and left (M_L) breast masses is shown in Fig. 12a. The measurements were related by $M_L = 1.02\ M_R + 25.1$ ($r = 0.97$) where the slope and intercept of the linear correlation were 1.02 ± 0.06 and 25.1 ± 44.1, respectively. In addition, in a Bland–Altman plot, the mean differences between masses from the right and left breasts were -35.4 ± 119.5 g (Fig. 12b).

The accuracy of the dual-energy technique for measuring water and lipid density was determined using chemical analysis as the reference standard. The relation of dual-energy water density (W_{DE}) and measured water density from chemical analysis (W_{CA}) is shown in Fig. 13a. The measurements were related by $W_{DE} = 1.13\ W_{CA} + 3.55\%$ ($r = 0.99$). In dual-energy decomposition, a fraction of protein (58%) in breast tissue is assigned to water so dual-energy water density was also correlated

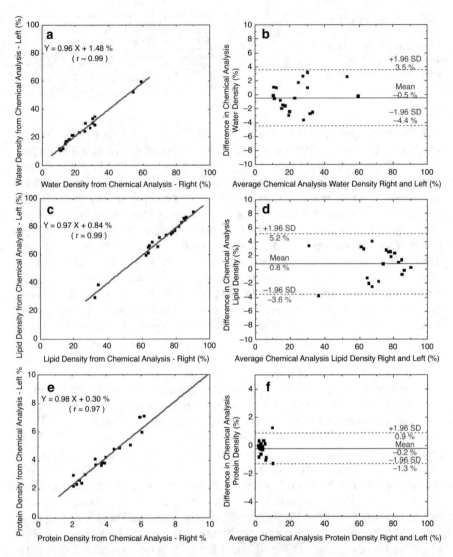

Fig. 11 Linear correlation of water (**a**) lipid (**c**) and protein (**e**) density measured with chemical analysis from the right and left breasts. In addition, in Bland–Altman plots, the mean differences between water, lipid and protein density from the right and left breasts were $-0.45 \pm 2.0\%$, $0.8 \pm 2.2\%$ and $-0.2 \pm 0.6\%$, respectively (**b, d, f**). SD standard deviation

with water density with the added protein from chemical analysis (WP_{CA}). The measurements were related by $W_{DE} = 1.13\ WP_{CA} + 0.94\%$ ($r = 0.99$). Therefore, the inclusion of protein to water did not change the correlation slope, but has shifted the offset toward zero. The relation of dual-energy lipid density (L_{DE}) and measured lipid density from chemical analysis (L_{CA}) is shown in Fig. 13c. The

Table 5 Summary of the linear regression analysis between right and left water, lipid, and protein from chemical analysis measurements

	Water	Lipid	Protein
Slope	0.96 ± 0.03	0.98 ± 0.03	0.98 ± 0.05
Intercept	$1.48 \pm 0.94\%$	$0.84 \pm 2.44\%$	$0.30 \pm 0.27\%$
Pearson's r	0.99	0.99	0.97
SEE	1.90%	2.15%	0.54%

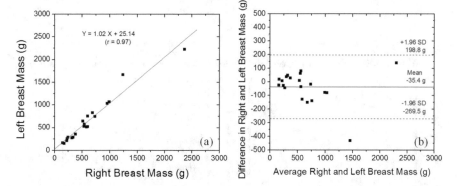

Fig. 12 Linear correlation of masses from the right and left breasts (**a**). In addition, in a Bland–Altman plot, the mean differences between masses from the right and left breasts were -35.4 ± 119.5 g (**b**). SD standard deviation

measurements were related by $L_{DE} = 0.91 \, L_{CA} + 0.24\%$ ($r = 0.99$). In addition, in Bland–Altman plots, the mean differences between water and lipid density from dual-energy mammography and chemical analysis were $4.5 \pm 2.7\%$ and $-6.4 \pm 2.5\%$, respectively (Fig. 13b, d). A summary of the linear regression analysis of water and lipid density from dual-energy and chemical analysis measurements is shown in Table 6 The water and lipid density from dual-energy mammography and chemical analysis were highly correlated with a standard error of estimate of 2.09% and 2.09%, respectively.

5.4 Discussions

Dual-energy mammography can be used to decompose only two different materials. It is possible to use water and lipid as basis materials given that they have the highest volumetric fraction in breast tissue. This will introduce a systematic error in estimation of water and lipid density. Our results indicate that the systematic error is relatively small due to the fact that the volumetric fraction of protein in breast tissue is on average 5.8% [40, 49].

In this study, the estimate of precision of water and lipid density for dual-energy mammography was determined by comparing the water and lipid density from two different views. The SEE for water and lipid density from two views

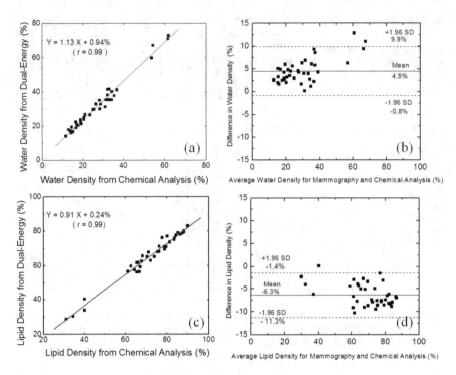

Fig. 13 Linear correlation of water (**a**) and lipid (**c**) density from dual-energy mammography and chemical analysis as the reference standard. In addition, in Bland–Altman plots, the mean differences between water and lipid density from mammography and chemical analysis were $4.5 \pm 2.7\%$ and $-6.4 \pm 2.5\%$, respectively (**b**, **d**). SD standard deviation

Table 6 Summary of the linear regression analysis of water and lipid density from dual-energy and chemical analysis measurements

	Water	Lipid
Slope	1.13 ± 0.03	0.91 ± 0.02
Intercept	$0.94 \pm 0.79\%$	$0.24 \pm 1.72\%$
Pearson's r	0.99	0.99
SEE	2.09%	2.09%

were calculated to be 1.25% and 1.11%, respectively. An estimate of precision for dual-energy mammography was also determined by comparing the water and lipid density from contralateral breasts, which also includes the biological variance between the contralateral breasts. The SEE for water and lipid density from the contralateral breasts with dual-energy mammography were calculated to be 2.81% and 2.74%, respectively. Furthermore, a comparison of water, lipid, and protein from right and left breasts using chemical analysis was also made. The SEE for water, lipid, and protein density from the right and left breasts with chemical analysis were calculated to be 1.90%, 2.15%, and 0.54%, respectively. These results indicate that the expected biological variance between the right and left breast is relatively small. Therefore, as previously reported [46], comparison of right and left breast density could be used as a practical approach to assess estimate of precision for breast density measurement techniques in an in vivo setting.

The accuracy of water and lipid density for dual-energy mammography was validated using chemical analysis in postmortem breasts. Water- and adipose-equivalent phantoms were used for dual-energy calibration [28]. Chemical analysis was used to measure the water, lipid, and protein components of breast tissue, and it was used as the reference standard to determine the accuracy of water and lipid density from dual-energy mammography. The SEE for water and lipid density from dual-energy mammography with respect to chemical analysis were calculated to be 2.09% and 2.09%, respectively. A previous study has reported a SEE of 4.6% for breast density measurement using dual-energy mammography as compared with chemical analysis [24], which is slightly higher than the SEE for water density from this study. However, the previous study showed a factor of 2 systematic error between breast density and volumetric glandular ratio from chemical analysis [24]. In this study, we have converted the water- and adipose-equivalent calibration phantoms into true water and lipid using a previously reported conversion matrix [28]. This method has significantly reduced the systematic error in water density. The mean difference between water density from dual-energy mammography and chemical analysis was estimated to be $6.8 \pm 2.7\%$. Furthermore, we have taken into consideration that approximately 58% of the protein content were decomposed into water in dual-energy decomposition [28]. Combining such fraction of protein with water from chemical analysis further reduced the systematic error to $4.5 \pm 2.7\%$, which is substantially lower as compared with the previous study [24]. In the linear regression analysis between water density measured from dual-energy mammography and chemical analysis, the slope was not affected by adding the protein fraction, but the offset was reduced from 3.55% to 0.94%. This is mostly due to the fact that protein fraction is positively correlated to water faction with a Pearson's r of 0.88 in this study.

Previous studies have also validated breast density measurement with MRI [49] and single-energy cone-beam CT [40] using standard image segmentation methods. A previous study has reported a SEE of 6.5% for breast density measurement using MRI as compared with chemical analysis [49]. Another study has reported a SEE of 3.9% for breast density measurement using CT as compared with chemical analysis [40]. Therefore, the SEE from dual-energy mammography was slightly higher than CT and lower than MRI.

In this study, breast density measurement using dual-energy mammography was implemented with an energy-sensitive photon-counting detection system [39, 50]. Dual-energy mammography has also been implemented using standard mammography with flat panel [24, 28, 46, 51, 52]. Therefore, clinical implementation of breast density measurement using dual-energy mammography is feasible. Dual-energy mammography provides higher accuracy and precision in breast density measurement as compared with other existing techniques [24, 28, 46, 51, 52]. Energy-sensitive photon-counting detection systems provide dual-energy capability without additional radiation dose. However, dual-energy mammography can also be implemented in standard flat panel-based systems with only minimal additional radiation dose [24, 28, 46, 51, 52].

A reliable and accurate breast density quantification technique can have significant impacts in current clinical practice. One of the potential applications is to use quantitative breast density measurement as an early assessment of therapeutic response to adjuvant endocrine therapy (AET), which is an established treatment for women with estrogen receptor-positive breast cancer. Previous studies on tamoxifen treatment, which is a common AET option for premenopausal women, have suggested that changes in breast density at 12–18 months can be used as an excellent biomarker for predicting treatment efficacy. On average, the annual reduction of breast density induced by tamoxifen was estimated to be approximately 4.3% [53]. However, existing breast density measurement methods are not reproducible enough to monitor small changes in breast density for individual women taking chemoprevention. In addition, to date, all clinical trials have evaluated changes in breast density based on visual assessment, which has substantial variability. Breast density measurement with dual-energy mammography is fully automated, which eliminates any observer variability. It is expected that the superior reproducibility of measuring breast density with dual-energy mammography will be sufficient to monitor changes in breast density within 1 year of endocrine therapy and will offer a more reliable approach for classifying women who are unlikely to benefit from endocrine therapy.

6 Conclusions

The establishment of the dual-energy technique for accurate breast density quantification allows for breast density to be better used in assessing breast cancer risk in patients. To date, methods to quantify breast density, including the current clinical standard using BI-RADS, have lacked the accuracy and precision needed to stratify women according to their risk. Previous studies show a twofold improvement in reproducibility for quantifying breast density using dual-energy mammography as compared with existing techniques [24, 46]. Therefore, dual-energy mammography can potentially address the impact of breast density as a risk factor, as it can largely eliminate variability in measuring breast density. The results from the postmortem study indicate that dual-energy mammography with water and lipid basis material calibration can be used to accurately measure volumetric water and lipid density in breast tissue. Improved quantification of breast tissue composition is expected to enhance its utility for assessment of response to therapy and as a risk factor for breast cancer.

Acknowledgments The authors would like to thank Dr. Hyo-Min Cho, David Sennung, Nikita Kumar, and Drs. Bahman Sadeghi, Hanna Javan, and Alfonso Lam Ng for their support in experiment and data analysis. The authors would also like to thank Dr. Erik Fredenberg for valuable technical support and acknowledge Philips Medical Systems for providing the MicroDose mammography system for this research. This work was supported in part by NIH/NCI grant R01CA13687.

References

1. Zhou P, Li YF, Harrison BL, Zhang GM, Smith D, Kelly MG, Schechter L, Robichaud AJ. MEDI 41-Novel 1-(arylsulfonyl)-3-(piperidinylmethyl)-1H-indoles as potent and selective 5-HT6 antagonists. Abstr Pap Am Chem Soc. 2007;234

2. Boyd NF, Byng JW, Jong RA, Fishell EK, Little LE, Miller AB, Lockwood GA, Tritchler DL, Yaffe MJ. Quantitative classification of mammographic densities and breast-cancer risk – results from the Canadian National Breast Screening Study. J Natl Cancer. 1995;I(87):670–5.

3. Widstrand C, Boyd B, Billing J, Rees A. Efficient extraction of toxic compounds from complex matrices using molecularly imprinted polymers. Am Lab. 2007;39:23–4.

4. Gilman BH, Kautter J. Consumer response to dual incentives under multitiered prescription drug formularies. Am J Manag Care. 2007;13:353–9.

5. Buist DS, Porter PL, Lehman C, Taplin SH, White E. Factors contributing to mammography failure in women aged 40–49 years. J Natl Cancer Inst. 2004;96:1432–40.

6. Byrne C, Schairer C, Wolfe J, Parekh N, Salane M, Brinton LA, Hoover R, Haile R. Mammographic features and breast-cancer risk – effects with time, age, and menopause status. J Natl Cancer. 1995;I(87):1622–9.

7. Carney PA, Miglioretti DL, Yankaskas BC, Kerlikowske K, Rosenberg R, Rutter CM, Geller BM, Abraham LA, Taplin SH, Dignan M, Cutter G, Ballard-Barbash R. Individual and combined effects of age, breast density, and hormone replacement therapy use on the accuracy of screening mammography. Ann Intern Med. 2003;138:168–75.

8. D'Orsi CJ, Mendelson EB, Ikeda DM. Breast imaging reporting and data system (BI-RADS). Reston: American College of Radiology; 2004.

9. D'Orsi CJ, Sickles EA, Mendelson EB, Morris EA. Breast imaging reporting and data system (BI-RADS). Reston: American College of Radiology; 2013.

10. Jackson VP, Hendrick RE, Feig SA, Kopans DB. Imaging of the radiographically dense breast. Radiology. 1993;188:297–301.

11. Kerlikowske K, Grady D, Barclay J, Sickles EA, Ernster V. Effect of age, breast density, and family history on the sensitivity of first screening mammography. JAMA. 1996;276:33–8.

12. Vacek PM, Geller BM. A prospective study of breast cancer risk using routine mammographic breast density measurements. Cancer Epidemiol Biomark Prev. 2004;13:715–22.

13. Berg WA, Campassi C, Langenberg P, Sexton MJ. Breast imaging reporting and data system: inter- and intraobserver variability in feature analysis and final assessment. AJR Am J Roentgenol. 2000;174:1769–77.

14. Ooms EA, Zonderland HM, Eijkemans MJC, Kriege M, Delavary BM, Burger CW, Ansink AC. Mammography: interobserver variability in breast density assessment. Breast. 2007;16:568–76.

15. Oza AM, Boyd NF. Mammographic parenchymal patterns – a marker of breast-cancer risk. Epidemiol Rev. 1993;15:196–208.

16. Young AM, McCabe SE, Boyd CJ. Adolescents' sexual inferences about girls who consume alcohol. Psychol Women Q. 2007;31:229–40.

17. Sivaramakrishna R, Obuchowski NA, Chilcote WA, Powell KA. Automatic segmentation of mammographic density. Acad Radiol. 2001;8:250–6.

18. Carton AK, Li JJ, Chen S, Conant E, Maidment ADA. Optimization of contrast-enhanced digital breast tomosynthesis. In: Astley SM, Brady M, Rose C, Zwiggelaar R, editors. Digital mammography, proceedings, vol. 4046; 2006. p. 183–9.

19. David AL, Torondel B, Zachary I, Ramirez MJ, Buckley SM, Cook T, Boyd M, Rodeck CH, Martin J, Peebles D. Local delivery of adenovirus VEGF to the uterine arteries increases vessel relaxation and uterine artery blood flow in the sheep. BJOG Int J Obstet Gynaecol. 2007;114:1034.

20. Lauder I, Holland D, Gowland G, Mason DY. Large cell tumors in lymph-nodes – immuno-histochemical identification using leukocyte common and prekeratin antibodies. J Pathol. 1983;140:164.

21. Burken JG, Sheehan E, Boyd M, Mayer P, Karlson U, Legind C. ENVR 179-in-planta sampling for site characterization. Abstr Pap Am Chem Soc. 2007;234.
22. Highnam R, Jeffreys M, McCormack V, Warren R, Smith GD, Brady M. Comparing measurements of breast density. Phys Med Biol. 2007;52:5881–95.
23. Cardinal HN, Fenster A. An accurate method for direct dual-energy calibration and decomposition. Med Phys. 1990;17:327–41.
24. Molloi S, Ducote JL, Ding H, Feig SA. Postmortem validation of breast density using dual-energy mammography. Med Phys. 2014;41:081917.
25. Rigie DS, La Riviere PJ. Optimizing spectral CT parameters for material classification tasks. Phys Med Biol. 2016;61:4599–623.
26. Jin KN, De Cecco CN, Caruso D, Tesche C, Spandorfer A, Varga-Szemes A, Schoepf UJ. Myocardial perfusion imaging with dual energy CT. Eur J Radiol. 2016;85:1914–21.
27. Ding HJ, Ducote JL, Molloi S. Breast composition measurement with a cadmium-zinc-telluride based spectral computed tomography system. Med Phys. 2012;39:1289–97.
28. Cho HM, Ding H, Kumar N, Sennung D, Molloi S. Calibration phantoms for accurate water and lipid density quantification using dual energy mammography. Phys Med Biol. 2017;62:4589–603.
29. Ducote JL, Klopfer MJ, Molloi S. Volumetric lean percentage measurement using dual energy mammography. Med Phys. 2011;38:4498–504.
30. Ducote JL, Molloi S. Quantification of breast density with dual energy mammography: an experimental feasibility study. Med Phys. 2010;37:793–801.
31. Ding H, Molloi S. Quantification of breast density with spectral mammography based on a scanned multi-slit photon-counting detector: a feasibility study. Phys Med Biol. 2012;57:4719.
32. Lam AR, Ding H, Molloi S. Quantification of breast density using dual-energy mammography with liquid phantom calibration. Phys Med Biol. 2014;59:3985.
33. Seo Y, Wong KH, Hasegawa BH. Calculation and validation of the use of effective attenuation coefficient for attenuation correction in In-111 SPECT. Med Phys. 2005;32:3628–35.
34. Boone JM, Seibert JA. An accurate method for computer-generating tungsten anode x-ray spectra from 30 to 140 kV. Med Phys. 1997;24:1661–70.
35. Ducote JL, Molloi S. Quantification of breast density with dual energy mammography: a simulation study. Med Phys. 2008;35:5411–8.
36. Ding H, Ducote JL, Molloi S. Measurement of breast tissue composition with dual energy cone-beam computed tomography: a postmortem study. Med Phys. 2013;40:061902.
37. Åslund M, Cederström B, Lundqvist M, Danielsson M. Scatter rejection in multislit digital mammography. Med Phys. 2006;33:933–40.
38. Fredenberg E, Hemmendorff M, Cederstrom B, Aslund M, Danielsson M. Contrast-enhanced spectral mammography with a photon-counting detector. Med Phys. 2010;37:2017–29.
39. Fredenberg E, Lundqvist M, Cederström B, Åslund M, Danielsson M. Energy resolution of a photon-counting silicon strip detector. Nucl Instrum Methods Phys Res, Sect A Accel Spectrom Detect Assoc Equip. 2010;613:156–62.
40. Johnson T, Ding H, Le HQ, Ducote JL, Molloi S. Breast density quantification with cone-beam CT: a post-mortem study. Phys Med Biol. 2013;58:8573–91.
41. Alvarez RE, Macovski A. Energy-selective reconstructions in x-ray computerised tomography. Phys Med Biol. 1976;21:733.
42. Brody WR, Butt G, Hall A, Macovski A. A method for selective tissue and bone visualization using dual energy scanned projection radiography. Med Phys. 1981;8:353–7.
43. Lehmann L, Alvarez RE, Macovski A, Brody WR, Pelc NJ, Riederer SJ, Hall AL. Generalized image combinations in dual KVP digital radiography. Med Phys. 1981;8:659–67.
44. Fredenberg E, Kilburn-Toppin F, Willsher P, Moa E, Danielsson M, Dance DR, Young KC, Wallis MG. Measurement of breast-tissue x-ray attenuation by spectral mammography: solid lesions. Phys Med Biol. 2016;61:2595.
45. Fredenberg E, Dance DR, Willsher P, Moa E, von Tiedemann M, Young KC, Wallis MG. Measurement of breast-tissue x-ray attenuation by spectral mammography: first results on cyst fluid. Phys Med Biol. 2013;58:8609.

46. Molloi S, Ding H, Feig S. Breast density evaluation using spectral mammography, radiologist reader assessment, and segmentation techniques: a retrospective study based on left and right breast comparison. Acad Radiol. 2015;22:1052–9.
47. Malkov S, Wang J, Kerlikowske K, Cummings SR, Shepherd JA. Single x-ray absorptiometry method for the quantitative mammographic measure of fibroglandular tissue volume. Med Phys. 2009;36:5525–36.
48. Kallenberg MG, van Gils CH, Lokate M, den Heeten GJ, Karssemeijer N. Effect of compression paddle tilt correction on volumetric breast density estimation. Phys Med Biol. 2012;57:5155–68.
49. Ding H, Johnson T, Lin M, Le HQ, Ducote JL, Su MY, Molloi S. Breast density quantification using magnetic resonance imaging (MRI) with bias field correction: a postmortem study. Med Phys. 2013;40:122305.
50. Aslund M, Cederstrom B, Lundqvist M, Danielsson M. Physical characterization of a scanning photon counting digital mammography system based on Si-strip detectors. Med Phys. 2007;34:1918–25.
51. Shepherd JA, Malkov S, Fan B, Laidevant A, Novotny R, Maskarinec G. Breast density assessment in adolescent girls using dual-energy X-ray absorptiometry: a feasibility study. Cancer Epidemiol Biomark Prev. 2008;17:1709–13.
52. Hodges NA, Tonjanika B, Han S. AGRO 60-the antioxidants in natural plants affecting formation of alpha synuclein nanobioparticles. Abstr Pap Am Chem Soc. 2007;234
53. Boyd NF, Lockwood GA, Byng JW, Little LE, Yaffe MJ, Tritchler DL. The relationship of anthropometric measures to radiological features of the breast in premenopausal women. Br J Cancer. 1998;78:1233–8.

Part II
Image Reconstruction and Material Discrimination

An Overview of CT Reconstruction with Applications to Photon Counting Detectors

Scott Hsieh

Half a century after its first introduction, x-ray computed tomography (CT) has now become a mainstay of radiology and medical diagnosis [1]. When it first debuted, the enabling technology behind CT was the computer, which at the time was a new device that allowed reconstruction of cross-sectional images. Today, CT is being further advanced by the introduction of the photon counting detector (PCD).

While the theory of CT reconstruction has been described in detail in the form of textbooks [2] and various review articles, the basic concepts of reconstruction are sometimes mysterious to developers of PCD technology. The goal of this chapter is to demystify some elements of CT reconstruction by providing a brief tutorial and overview, so that detector subsystem experts such as ASIC designers and high-Z semiconductor specialists can better understand the systems trade-offs that they face. We seek to provide a working understanding only, not a complete mathematical derivation. The discussion will be centered on new applications provided by PCD technology.

This chapter is organized as follows. First, we will provide a working overview of CT reconstruction, starting from the foundational projection-slice theorem and ending with a qualitative understanding of helical reconstruction, iterative reconstruction, and deep learning denoising. We will especially elaborate on the higher spatial resolution of PCD CT and how this affects reconstruction. Second, we will describe spectral reconstruction and how multibin data are transformed into spectral images. We will close with some practical comments on how PCD data would be used in practice and how knowledge of CT reconstruction could affect PCD subsystem design.

S. Hsieh (✉)
Mayo Clinic, Rochester, MN, USA
e-mail: hsieh.scott@mayo.edu

© The Author(s), under exclusive license to Springer Nature Switzerland AG 2023
S. Hsieh, K. (Kris) Iniewski (eds.), *Photon Counting Computed Tomography*,
https://doi.org/10.1007/978-3-031-26062-9_7

1   Reconstruction Fundamentals

CT, as an imaging modality, can be decomposed into three parts: (1) an x-ray system including source and detector, (2) a rotating gantry that allows the x-ray system to acquire multiple views from different angles, and (3) a computer that implements reconstruction. Reconstruction itself is the process wherein raw x-ray data, sometimes called the sinogram, is transformed into cross-sectional images.

The projection-slice theorem   Simplifying to a 2D environment, the CT scanner, over the course of a complete rotation, samples every possible line through the object. In most cases, the patient fits entirely within the scanner; if not, truncation artifacts result, which is outside the scope of the tutorial. Using the Beer–Lambert law, the total line integral of attenuation $\mu(x, y)$ through the path of the x-ray beam can be obtained as follows:

$$I = I_0 e^{-\int_r \mu(x,y)dr}$$

where I_0 is the intensity of the incident x-ray beam, I is the intensity of the detected x-ray beam, and the integral is taken over the line connecting the source to the detector pixel. Strictly speaking, this equation is only valid for monoenergetic x-rays without scatter. Polyenergetic radiation or scatter will cause artifacts which must be corrected in a separate processing step.

Figure 1 shows an example sinogram through the Shepp–Logan phantom. The sinogram measures all line integrals through the phantom. From the sinogram, the task of reconstruction is to recreate the original underlying object that was scanned.

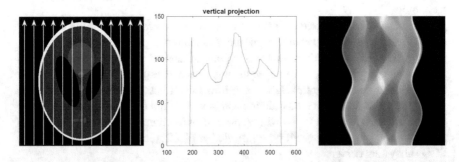

Fig. 1 (Left) The numerical Shepp–Logan phantom, loosely modeled after the head. Arrows correspond to direction of integration in the first projection. (Center) First said projection, traversing the same lines as the arrows. Peaks correspond to integration through bony anatomy or through hyperattenuating material in the center. (Right) Sinogram, composed of several views from several different directions. The first projection is the top row of the sinogram, and subsequent rows are from different angles. Reconstruction maps the sinogram back to an estimate of the original image

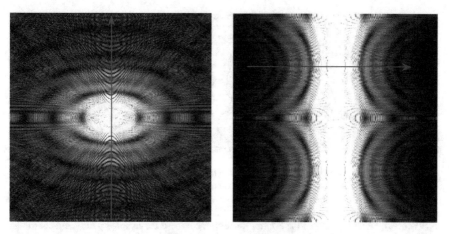

Fig. 2 The projection-slice theorem. (Left) Two-dimensional Fourier transform of the Shepp–Logan phantom, shown in Fig. 1. (Right) One-dimensional Fourier transform of the sinogram, previously shown also in Fig. 1. The Fourier transform is taken over the detector direction, not the gantry rotation direction. The red lines contain the same data and can be seen to have the same patterns. From the sinogram, one can fill out the Fourier transform of the object to be reconstructed, and then an inverse Fourier transform will yield the original object

The transformation between sinogram and reconstruction is best understood in the frequency (also called Fourier transform) domain. The projection-slice theorem (also called central slice theorem), first derived by Bracewell for the problem of radio astronomy [3], states that measurements of a set of parallel line integrals map to measurements of a corresponding line in the frequency domain. In the original first-generation CT scanners, these parallel measurements were acquired by translation of the source and single detector pixel. Today, a single projection samples a diverging set of line integrals from the source into a curved or linear detector, but the data from fan-beam CT can still be rebinned to create a parallel-beam CT dataset, and indeed some CT vendors prefer rebinning over direct fan-beam reconstruction because the noise and sampling properties become more uniform. Graphically, the projection-slice theorem is shown in Fig. 2.

Filtered backprojection and its relationship to the projection-slice theorem
Reconstruction can therefore be performed directly in the frequency domain. One difficulty is that the samples in frequency domain are not uniformly spaced on a Cartesian grid. It is nonetheless possible to use a gridding algorithm to resample the data onto a Cartesian grid, in which case a fast inverse Fourier transform would directly yield the reconstructed image. A nonuniform Fourier transform algorithm could also be used instead of gridding.

In practice, filtered backprojection is commonly used instead. In filtered backprojection, the set of parallel line integrals are first filtered by a ramp filter, and then they are smeared back along their direction of travel. It can be shown that filtered backprojection is mathematically equivalent to the gridding process: projecting a

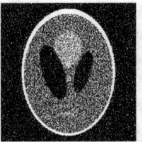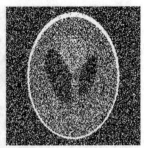

Fig. 3 The consequences of higher resolution. (Left) Sampling density in Fourier transform space. Following the projection-slice theorem, each projection provides data for one spoke in 2D Fourier transform space. The center of Fourier transform (low-frequency data) is well sampled, but the outer boundaries (high-frequency data) are not. When higher resolutions become available, we can now access a larger territory of the Fourier transform, but these are sampled poorly and contribute greater noise. (Center) Noisy reconstruction with three-pixel binning. (Right) Noisy reconstruction without pixel binning. Note that some low-contrast features are now difficult to see, even though the data is still present

single projection fills out the corresponding line in Fourier space, and the ramp filter compensates for the variable sampling density within Fourier space (if the ramp filter was not applied, low frequencies would be amplified). Advantages of filtered backprojection over gridding include a more natural extension to helical reconstruction, possible reduction of artifacts in gridding estimation, and a long history of implementation expertise.

Consequences of high-resolution CT A major consequence of this theoretical understanding of reconstruction is that the sampling density is greatest at low frequencies, because many "spokes" of the Fourier transform are redundantly measured. In fact, the density goes as $1/k$, where k is the magnitude of the frequency. Figure 3 shows the sampling density of the Fourier transform with a small number of views; note the higher density of lines at the center. For rays of uniform detected intensity and for pixels without crosstalk, the noise of the samples in frequency space is uniform and independent, so that changes in noise as a function of frequency are driven entirely by the sampling density. Therefore, higher frequencies have greater noise. In particular, when the spatial resolution is doubled, the amount of frequency space that is sampled is doubled along both the x-direction and the y-direction. For low frequencies, the measured signal and the noise are both unchanged when the detector pixel size decreases. However, higher resolutions give us *access* to higher spatial frequencies. These spatial frequencies have fewer samples and necessarily have greater noise. The variance of the reconstruction is proportional to the total noise power, and so the variance increases very quickly with resolution.

Figure 3 illustrates the consequences of smaller detector pixels. The center subpanel shows a reconstruction at lower resolution. The right subpanel shows the reconstruction at higher resolution. From an information perspective, the right subpanel is strictly better than the center subpanel, although it may not seem strictly advantageous to the human visual system. If a Fourier transform were taken from the center and right subpanel, we would see that they are exactly the same at low to medium frequencies, but beyond a cutoff frequency, the high-resolution detector has more data – data that is disproportionately corrupted by noise. For a computer detection algorithm, it is trivial to transform the right, high-resolution panel to the center, low-resolution subpanel by applying the Fourier transform and then a cutoff filter, so detection at high resolution is strictly better than detection at low resolution. However, the human visual system is not optimized for the colored noise characteristics of CT. The human visual system can adequately compensate for white noise, but not the colored noise that results from ramp-spectrum filtration, so the low-contrast targets at the bottom of the Shepp–Logan phantom become more difficult to perceive with the high-resolution reconstruction. For this reason, apodization – or simply, "blurring" – is used to reduce noise power, as well as signal power, at higher frequencies. At some point, the PCD might as well bin the detector data first. After all, binning the data reduces strain on the data transmission system.

In some applications, high-frequency information is essential and the signal-to-noise ratio is high. One example is in examination of the temporal bone [4]. For this application, PCD CT can be very useful and reduce dose by over 80%. It is noteworthy that in some previous systems, high resolutions were created using a comb filter to eliminate half the signal, already transmitted through the patient, before it reaches the detector. Compared to this option, the high resolution of PCD CT is very advantageous for reducing dose.

Data bandwidth for high-resolution CT Although not the main objective of this chapter, we comment briefly on the difficulty of transmitting large amounts of data in CT. PCD CT scanners collect and transmit much more data than their energy-integrating counterparts. With a pixel density approximately $10\times$ greater than conventional detectors (assuming, e.g., a typical pixel pitch of 0.33 mm for PCD and 1.0 mm for conventional detectors), assuming that higher spatial resolution demands $2\times$ denser samples in the gantry rotation direction to reduce viewstreak artifacts, and assuming four energy bins, we find that the PCD CT demands 80 times the data bandwidth of conventional CT. This huge amount of data bandwidth is a challenge, especially because standard wired connections are not feasible with a continuously rotating gantry: data must be transmitted either through slip-ring contact or wirelessly.

It should be pointed out that this is not the first time CT has faced this conundrum. In the early days of multi-slice CT, data bandwidth was $4\times$–$16\times$ more than prior generations. Some systems used adaptive binning, so that, for example, only four slices were transmitted even when 16 physical detector rows existed. The collimation and binning were configured so that they could either be thin slices (eliminating the outer 12 slices, leading to high resolution with limited coverage) or thick slices (binning together groups of four detector rows, leading to low axial resolution but greater coverage), depending on the clinical application. A similar strategy is being pursued by PCD CT vendors today, with adaptive binning used so that high resolution is only pursued in select applications, perhaps with trade-offs such as reduced coverage or slower gantry rotation time.

Other strategies for reducing data bandwidth are possible. Binning in the spectral direction has room for improvement. Even if ten energy bins are collected, these are redundant measurements of only two basis materials, so a data reduction step is theoretically possible on-chip prior to slip-ring transmission that should still be noise efficient, as we will discuss later. In the spatial direction, it may be possible to practice lossy compression in a way that is more efficient than direct binning.

Non-idealities in reconstruction While the Beer–Lambert law and the projection-slice theorem form the theoretical basis for reconstruction, improvements and refinements are routinely applied on modern CT scanners. First, the Beer–Lambert law is itself an approximation of reality. Real measurements are polyenergetic, not monoenergetic as the Beer–Lambert law presumes, and there is scatter that contaminates and adds signal to the final measurements. To accommodate beam hardening in objects of a single material, a calibration can be used. Typically, a polynomial correction is sufficient to linearize the response. With multiple materials, an approach that is effective in conventional CT is an iterative algorithm that first detects bone or other materials within the scan and then uses estimates of path lengths through normal tissue and through bone to guide a calibration [5]. In many cases, a single iteration is sufficient. In PCD CT, it is possible to improve upon this by using spectral measurements to guide beam hardening correction. For more information on this topic, Kimoto et al. describe in this book an algorithm to generate effective atomic number images using polynomial linearizations.

A separate problem in commercial CT scanners is imperfection of individual detector pixels. In the early days of fan-beam CT, detector drift was a substantial problem, and many detector subsystems were rejected as being non-conformant to quality control standards. Small changes in detector calibration, much less than 1%, can lead to large ring artifacts in reconstruction both because CT is sensitive to small changes in tissue (with 1 HU being 0.1% the density of water) and because of the ramp filter applied during filtered backprojection, which amplifies differences between neighboring detector pixels. Since then, hardware manufacturing and software correction algorithms have both improved. PCD CT is still nascent and the manufacturing processes have not yet matured. The use of aggressive ring artifact corrections may be necessary in the early years of PCD CT to generate acceptable images, and these algorithms do exist [6]. The basic principle of ring artifact correction is to detect pixels that are consistently brighter or darker than neighbors and to apply an ad hoc correction to equalize them. If the root cause of this drift can be identified, smarter and more effective corrections might be applied. For example, continued illumination through air of some pixels could lead to polarization and may create artifacts analogous to detector lag faced in other systems. In particular, cone-beam CT detectors, often amorphous silicon panels, have faced similar problems in calibration reliability and generate both ring artifacts and detector lag artifacts. These systems are traditionally not expected to provide the same level of fidelity as diagnostic CT systems (e.g., they may be used for bony landmark alignment in radiotherapy rather than for detection of low-contrast metastases), but PCD CT systems will be expected to fulfill the same demands as traditional CT and must fulfill more exacting calibration specifications.

Helical reconstruction and iterative reconstruction Two major advances have occurred in reconstruction that have made their way into clinical systems. The first is helical reconstruction, which arose from the desire to create higher-resolution images in the z-direction: today, CT features isotropic resolution with the ability to create high-fidelity multiplanar reformats, but prior to multi-slice CT, the resolution in these dimensions was approximately 1 cm (depending on collimation and time spent scanning). Helical reconstruction has been empirically developed. The popular "FDK" or Feldkamp cone-beam CT algorithm is essentially an extrapolation from direct fan-beam reconstruction. A mathematically exact helical reconstruction took years to develop [7], but empirically developed algorithms have better noise statistics and sampling. It is accurate to say that helical reconstruction today is largely an extension of 2D reconstruction.

Iterative reconstruction [8] has now been in place for about 15 years, and while the details of fast commercial implementations are not known to us, the basic principle of the model-based iterative reconstructions is to apply a regularizer that penalizes small differences between neighboring voxels, which are likely to result from noise, while not penalizing large differences that likely represent edges. These reconstruction algorithms could be viewed as an intelligent, adaptive apodization across the reconstructed volume. There are other advantages also such as better modeling of the ray sampling geometry, but these are expected to be minor improvements except in high-resolution tasks.

Deep learning reconstruction Newer, AI-based reconstruction algorithms are now starting to become available. These reconstruction algorithms function using a large dataset of matched less noisy and more noisy data, with the more noisy data used as training input and the less noisy data used as training output. These matched datasets are fed into convolutional neural networks. While it is tempting to imagine that large datasets can yield dramatic improvements in image quality and dose reduction, based on the apparent image quality of the generated images, our view is that deep learning reconstruction only functions as a further refinement of the apodization presented earlier in Fig. 3 (compensating for the inability of the human visual system to deal with noise at high spatial frequencies) or the regularizer used in iterative reconstruction, which can be understood to adapt the apodization locally to the noise or signal texture. Overly aggressive applications of deep learning reconstruction could cause some lesions to disappear. This is not to say that these algorithms have no utility; in contrast, they can dramatically simplify interpretation, especially for tasks with high contrast such as detection in the lung. However, the level of dose reduction that can be applied is not arbitrarily large. Because they are computationally efficient and can compensate for high-frequency noise, they are an attractive option for PCD CT.

2 Spectral Reconstruction

Besides the increased resolution discussed using Fig. 3, the primary advantage of PCD is the universal availability of spectral images. Spectral CT images are associated with several clinical applications [9]. However, it should be emphasized that in most applications, the primary use of CT is in grayscale or conventional imaging, with spectral CT adding incremental value depending on the specific clinical indication. In a few cases, spectral CT can drive the indication, but in most cases, it will only be used on an incidental basis.

The difference between "spectral CT" and "dual-energy CT" is that spectral CT involves acquisition of two or more energies, whereas dual-energy CT is limited to two energies. In many applications, the differences between two and more than two energies are limited to improvement of noise. Unless k-edge contrast materials are available, the two fundamental contrast mechanisms present in CT are photoelectric effect and Compton scattering; however, for an alternative viewpoint on this question, the interested reader can consult with Xiangyang Tang's chapter in this book. Regardless, it is safe to say that the interactions of x-rays with the human body in the diagnostic energy range are well approximated by decomposing the body into photoelectric and Compton components, as was known early in the days of CT [10, 11]. The Beer–Lambert law with multiple energy measurements can be transformed into a system of linear equations, from which one can solve for the photoelectric and Compton components separately. These can be reconstructed to produce photoelectric and Compton basis material images. A virtual monoenergetic image (VMI) can then be created as a linear sum of the photoelectric and Compton basis images, being equivalent to an image that would be generated from a hypothetical monoenergetic x-ray source. Having more than two energies does not enable reconstruction of another basis material unless a k-edge basis material is present. In most clinical applications, the k-edge of iodine is not adequately high to provide a third basis material, as there is little penetration below the iodine k-edge. However, more energy bins reduces the noise of the spectral images by providing some level of redundant measurements that can be averaged together [12].

It is not necessary to use PCDs for spectral imaging. Spectral imaging has previously been available on flagship systems from several vendors, but each system has its trade-offs. In the dual-source implementation, two sources and two detectors are necessary, roughly doubling the hardware requirements of the scanner. However, the dual-source implementation allows the application of tin filtration, improving the spectral separation. In the fast-kVp switching implementation, the x-ray source alternates between low and high kV (e.g., 80 and 140 kV) in every other projection. This allows near-simultaneous acquisition of both energy datasets with excellent co-registration, but because it is not possible to filter one energy only without a complex switching mechanical apparatus, the energy separation is inferior. A limitation of both these implementations is that a dedicated dual-energy acquisition protocol must be used. This is acceptable for certain clinical applications, like kidney stone characterization [13] or measurement of gout [14], but it is not suited for incidental findings like the characterizing of ambiguous renal lesions [15]. The dual-layer detector implementation provides universal spectral imaging, useful for retrospectively analyzing incidental findings, but its spectral separation is poor.

The requirement in spectral imaging, then, is to recover the underlying basis materials. Once the basis materials are recovered, reconstruction can proceed as usual: first for the Compton basis material, and then for the photoelectric basis material, and then combinations of these can be used for grayscale CT tasks as virtual monoenergetic images. The process of transforming pixel energy bin measurements to basis material measurements is sometimes called "basis material estimation" or simply" estimation," and the algorithm implementing this process is the estimator algorithm.

There are two possible conceptual frameworks for basis material estimation. One approach, favored by some physicists, is to presuppose some exact mapping between the pixel energy bin measurements and the underlying basis material quantities. The second approach, favored by some practitioners, is to empirically identify a transformation from energy bin measurements to approximately recover basis materials first. Differences between these approaches lay in the kinds of non-idealities that must be addressed in the mind of the implementer. To the physicist, non-idealities such as scatter, pixel drift, or imperfect calibration are nuisances for which better engineering, improved anti-scatter grids, and the march of time will inevitably remedy. To the practitioner, today's photon counting detectors are by their nature imperfect but clinical products must be made from them. It is therefore imperative that an algorithm be developed that produces immediately acceptable images, regardless of the theoretical justification. Small noise improvements that are attractive to the physicist may be of secondary importance to the practitioner who simply wants a first acceptable clinical product. We will first discuss the viewpoint of the physicist, and then we will discuss the viewpoint of the practitioner.

The physicist seeks the optimal solution to basis material estimation. When there are as many energy bin measurements as basis materials (e.g., two of each), then the mapping is necessarily one-to-one. When the number of energy bins exceeds the number of basis materials, the solution is overdetermined. When four energy bins are available, the problem can be simplified by binning data into two bins, but this is associated with increases in noise because information is being thrown away. However, the mapping between energy bins to basis materials is smooth, and by linearizing the system and then applying a bias correction, nearly optimal performance could be achieved [16]. The accuracy of linearization decreases at different flux regimes so multiple linearization options can be used instead [17]. An important tool to the physicist is the Cramér–Rao lower bound of the variance, which specifies the best possible variance that can be achieved by an unbiased estimator [18].

The practitioner may be inclined to view the theoretical decomposition of energy bin measurements into basis materials with a dose of skepticism. Beam harden-ing, scatter, and pixel imperfections play heavily in spectral imaging. Excellent reconstruction algorithms are available for classical grayscale CT, going beyond simple filtered backprojection by applying a variety of hand-tuned noise reduction and artifact mitigation options. These noise reduction tools are often not based in theory but in experience, with parameters tuned to create visually pleasing images with noise texture that is acceptable to radiologists. Furthermore, the basis material decomposition proposed by the physicist produces very noisy images, even when achieving the so-called maximum performance of the Cramér–Rao lower bound, because the noise reduction tools that are long familiar to the practitioner are not used. Rather than achieving theoretically maximum noise performance and visually unappealing images, the practitioner may rely on other noise reduction algorithms that create visually acceptable images that are already in use, even if the underlying signal-to-noise ratio is slightly compromised, because the final product may be more acceptable to the radiologist. A major observation here for noise reduction in spectral images is that the errors in the two basis materials are strongly anticorrelated; regularization that exploits this anticorrelation creates images that are visually more pleasing, and probably higher performance on human detection experiments, although this supposition has not been tested [19].

Therefore, the practitioner may be inclined to format or transform the sinogram so that it is more similar to conventional CT and then to apply the familiar toolbox of algorithms to reduce noise or suppress scatter. The spectral component can then be integrated later for various corrections, including beam hardening artifacts. Rather than work through the mathematics of spectral estimation, the practitioner can directly appeal to the CT number ratio, which is a substantially simpler construct.

Figure 4 illustrates the use of the CT number ratio for the application of kidney stone classification. To use the CT number ratio, the sinogram for each energy bin is independently formed and reconstructed. After the reconstruction, the ratio between the low and the high CT number of a lesion is calculated, and this ratio gives an estimate of the relative contribution of photoelectric and Compton components. The CT number ratio is easy to use and understand and does not require a spectral estimator algorithm. From Fig. 4, it is possible to appreciate what might happen if the spectral separation decreases. For example, if, instead of 80 and 140 kV, a dual-source system used 100 and 120 kV, it could be appreciated that the contrast between bone and gout would decrease, and the two ellipses merge closer together. In the case of conventional dual-energy imaging, the noise of the low- and high-energy CT numbers is independent and (to first approximation) similar regardless of the energy used to acquire the data. Therefore, it is desirable to use highly separated energies to maximize spectral contrast, except for large patients that may present too much attenuation at very low energies.

The picture is more complex, but analogous, for multibin PCDs. One obvious complexity is that there may be more than two energy bins, so it is not obvious how to directly calculate the "ratio" of more than two quantities efficiently. For example, for a four-bin PCD, one might average the two higher energies and average the two lower energies, but this will not be fully efficient. A separate complexity is that the energy bins are correlated: charge sharing, for example, may cause a high-energy photon to be detected as multiple low-energy events. Correcting for charge sharing would increase spectral separation but adds dependencies and may increase noise as well.

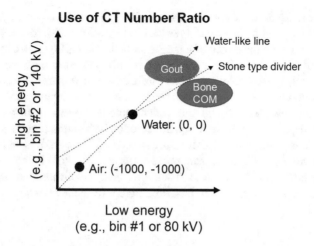

Fig. 4 Use of the CT number ratio, the simplest approach to using multi-energy or spectral data. Images are reconstructed using the lower energy and higher energy separately. For a lesion of interest, the CT numbers of both reconstructions are analyzed and compared. In this application of kidney stone classification, it is expected that gout (uric acid, similar atomic number to water) will have its low- and high-energy bin CT number be comparable, whereas for calcium monohydrate (COM) and similar stone types, the low-energy CT number will be much higher than the high energy. For kidney stone applications, a dividing line can be placed, and stones below the line will be classified as calcium-containing stones, whereas stones above the line will be classified as uric acid stones. In gout applications, the dividing line instead separates gout from pseudogout, a condition that appears similar to gout but with calcified deposits rather than uric acid deposits

In this landscape, PCD engineers may wonder how best to design their detectors and what aspects to prioritize. It should be emphasized that in these early days, stability and reproducibility are of paramount importance because they enable clinically reliable systems. However, as we move to second-generation PCDs, it is also important to improve the limiting performance of the detector. It is not necessary to fully implement the Cramér–Rao lower bound, which may be seen as an arcane mathematical construct. Rather, one can image a uniform object with small additional attenuators. For example, one could image a 20 cm acrylic block (as a model for human attenuation) and then apply 5 mm of additional acrylic in one strip, and 0.5 mm of aluminum in another strip. From the N energy bins, one can construct, using multivariable linear regression, a linear fit such that (1) the background signal, when imaging only acrylic, averages to zero; (2) the data behind the aluminum average to 0.5; (3) the data behind the additional acrylic averages to 0; and (4) the weights selected produce the smallest variance within one of the regions (e.g., the background signal region). This is effectively a least squares problem that creates an aluminum basis material estimate, centered at one operating point. Because we have specified a mixture of hard constraints together with a least squares variance minimization, this is a constrained least squares problem and can be solved appropriately. An analogous process can be performed for the acrylic material, to create an acrylic basis material. The extra machinery of the Cramér–Rao lower bound is not needed; in most cases, a simple least squares formulation achieves the bound and is easy to implement. If three basis materials are desired, the setup can be extended to include a third strip of material. In fact, it could be argued that the extra machinery of the Cramér–Rao lower bound is only present to have the appearance of a theoretically perfect solution. When carefully examined, the Cramér–Rao lower bound has two terms: one term which represents the least squares answer that we have discussed here and a second term which captures changes in the variance. This second term departs from the least squares solution but is commonly neglected in x-ray applications because it is usually close to zero and because it is cumbersome to measure accurately, needing large numbers of noise realizations before a sufficiently unbiased estimate can be achieved. Therefore, many applications of the Cramér–Rao lower bound in papers are, in fact, no different from the least squares approach outlined here.

3   Take-Home Messages: Reconstruction for Detector Scientists

As we close this chapter, let us review several specific messages relevant to PCD engineers and scientists wanting to understand the impact of reconstruction on their PCD design choices:

- Modern PCDs provide much better spatial resolution and always-available spectral imaging than their conventional, energy-integrating counterparts.
- Better spatial resolution provides access to higher frequencies in the two-dimensional Fourier transformation. These frequencies can be measured using high-resolution detectors, but because of the projection-slice theorem, sampling in this regime is fundamentally poor. Nothing can be done to remedy this. The reconstruction is therefore very noisy. We emphasize that nothing is lost by acquiring data at these high frequencies, but there are costs in data bandwidth and reconstruction time, and the human visual system is not efficient at using this information – detection by humans is better performed if apodization (or blurring) is first performed, assuming that the lesion is several millimeters in diameter or large and of low contrast (Fig. 3). If only larger lesions need to be detected and classified, the detector might as well be binned, as the apodization would effectively blur it to the same degree anyway during reconstruction.
- Theoretically, the quality of spectral imaging is best assessed using the Cramér–Rao lower bound of the variance. Practically, a good surrogate is basis material subtraction, which is a point estimate of this lower bound and can easily be experimentally measured at a specified operating point. With multiple energy bins, the Cramér–Rao lower bound is similar to basis material subtraction, where the weights are selected through solving a constrained least squares problem.
- The CT number ratio is a useful construct in that it reduces spectral reconstruction to the previously solved and optimized problem of single-energy

reconstruction. Alternatively, basis material projections may be produced first using basis material estimation, and these images can be reconstructed independently and then merged into virtual monoenergetic images. The latter approach more directly addresses fundamental physics limitations, but the former approach may be simpler to achieve in practice.

References

1. Boone JM, McCollough CH. Computed tomography turns 50. Phys Today. 2021;74(9):34–40.
2. Kak AC, Slaney M. Principles of computerized tomographic imaging. SIAM; 2001.
3. Bracewell RN. Strip integration in radio astronomy. Aust J Phys. 1956;9(2):198–217.
4. Rajendran K, Voss BA, Zhou W, Tao S, DeLone DR, Lane JI, et al. Dose reduction for sinus and temporal bone imaging using photon-counting detector CT with an additional tin filter. Investig Radiol. 2020;55(2):91.
5. Hsieh J, Molthen RC, Dawson CA, Johnson RH. An iterative approach to the beam hardening correction in cone beam CT. Med Phys. 2000;27(1):23–9.
6. Prell D, Kyriakou Y, Kalender WA. Comparison of ring artifact correction methods for flat-detector CT. Phys Med Biol. 2009;54(12):3881.
7. Katsevich A. Theoretically exact filtered backprojection-type inversion algorithm for spiral CT. SIAM J Appl Math. 2002;62(6):2012–26.
8. Thibault JB, Sauer KD, Bouman CA, Hsieh J. A three-dimensional statistical approach to improved image quality for multislice helical CT. Med Phys. 2007;34(11):4526–44.
9. Hsieh SS, Leng S, Rajendran K, Tao S, McCollough CH. Photon counting CT: clinical applications and future developments. IEEE Trans Radiat Plasma Med Sci. 2020;5(4):441–52.
10. Lehmann L, Alvarez R, Macovski A, Brody W, Pelc N, Riederer SJ, et al. Generalized image combinations in dual KVP digital radiography. Med Phys. 1981;8(5):659–67.
11. Alvarez RE, Macovski A. Energy-selective reconstructions in x-ray computerised tomography. Phys Med Biol. 1976;21(5):733.
12. Faby S, Kuchenbecker S, Sawall S, Simons D, Schlemmer HP, Lell M, et al. Performance of today's dual energy CT and future multi energy CT in virtual non-contrast imaging and in iodine quantification: a simulation study. Med Phys. 2015;42(7):4349–66.
13. Primak AN, Fletcher JG, Vrtiska TJ, Dzyubak OP, Lieske JC, Jackson ME, et al. Noninvasive differentiation of uric acid versus non–uric acid kidney stones using dual-energy CT. Acad Radiol. 2007;14(12):1441–7.
14. Glazebrook KN, Guimarães LS, Murthy NS, Black DF, Bongartz T, Manek NJ, et al. Identification of intraarticular and periarticular uric acid crystals with dual-energy CT: initial evaluation. Radiology. 2011;261(2):516–24.
15. Mileto A, Allen BC, Pietryga JA, Farjat AE, Zarzour JG, Bellini D, et al. Characterization of incidental renal mass with dual-energy CT: diagnostic accuracy of effective atomic number maps for discriminating nonenhancing cysts from enhancing masses. Am J Roentgenol. 2017;209(4):W221–W30.
16. Alvarez RE. Estimator for photon counting energy selective x-ray imaging with multibin pulse height analysis. Med Phys. 2011;38(5):2324–34.
17. Rajbhandary PL, Hsieh SS, Pelc NJ. Segmented targeted least squares estimator for material decomposition in multibin photon-counting detectors. J Med Imaging. 2017;4(2):023503.
18. Roessl E, Herrmann C. Cramér–Rao lower bound of basis image noise in multiple-energy x-ray imaging. Phys Med Biol. 2009;54(5):1307.
19. Zhang R, Thibault J-B, Bouman CA, Sauer KD, Hsieh J. Model-based iterative reconstruction for dual-energy X-ray CT using a joint quadratic likelihood model. IEEE Trans Med Imaging. 2013;33(1):117–34.

On the Choice of Base Materials for Alvarez–Macovski and DIRA Dual-energy Reconstruction Algorithms in CT

Maria Magnusson, Gudrun Alm Carlsson, Michael Sandborg, Åsa Carlsson Tedgren, and Alexandr Malusek

1 Introduction

The Alvarez–Macovski method (AM) [1] and the dual-energy iterative reconstruction algorithm (DIRA) [2] are image reconstruction algorithms in dual-energy computed tomography (DECT) using a mathematical decomposition of the energy-dependent linear attenuation coefficient (LAC) into energy-dependent basis functions. Both algorithms use energy spectra of photons emitted from the X-ray tube, and both algorithms can produce virtual monoenergetic images at any energy.

The choice of the basis functions affects the accuracy of the resulting weight coefficients and the reconstructed LAC. In the medical diagnostics energy range of 20–150 keV, the LAC at photon energy E can be written as a sum of the

M. Magnusson (✉)
Department of Electrical Engineering, Linköping University, Linköping, Sweden

Department of Health, Medicine and Caring Sciences, Linköping University, Linköping, Sweden

Center for Medical Image Science and Visualization (CMIV), Linköping University, Linköping, Sweden
e-mail: maria.magnusson@liu.se

G. A. Carlsson
Department of Health, Medicine and Caring Sciences, Linköping University, Linköping, Sweden

Center for Medical Image Science and Visualization (CMIV), Linköping University, Linköping, Sweden

M. Sandborg
Department of Health, Medicine and Caring Sciences, Linköping University, Linköping, Sweden

Center for Medical Image Science and Visualization (CMIV), Linköping University, Linköping, Sweden

© The Author(s), under exclusive license to Springer Nature Switzerland AG 2023
S. Hsieh, K. (Kris) Iniewski (eds.), *Photon Counting Computed Tomography*,
https://doi.org/10.1007/978-3-031-26062-9_8

photoelectric absorption, $\mu_p(E)$, incoherent scattering, $\mu_{inc}(E)$, and coherent scattering, $\mu_{coh}(E)$, components:

$$\mu(E) = \mu_p(E) + \mu_{inc}(E) + \mu_{coh}(E). \qquad (1)$$

Each of these components, also known as macroscopic cross sections, is related to the (microscopic) cross section, σ_i, for the corresponding interaction process i (p, inc or coh) as $\mu_i = n_V \sigma_i$, where n_V is the number of interaction centers (typically atoms) per unit volume; see Ref. [3] for more information. Unless mentioned otherwise, dependencies of the cross section on energy and the atomic number, Z, reported in this section are also taken from Ref. [3].

The photoelectric effect is when an atomic electron absorbs the incident photon. The process is more probable for tightly bound electrons in inner shells than weakly bound electrons in outer shells. The atomic photoelectric cross section for K-shell electrons at some distance from the K-shell is proportional to $Z^n/E^{3.5}$, where n ranges from 4 at low photon energies to 4.6 at high photon energies, and Z^5/E for relativistic energies. The $1/E^3$ dependence is often taken as a transition between the two cases. Theoretical predictions of μ_p near the absorption edge are difficult and uncertain.

The incoherent scattering is when an atomic electron scatters the incident photon, and a part of the photon's energy is transferred to the electron. The process was first described as a scattering of a photon on a free electron by Compton and so this interactions is sometimes referred to as the Compton scattering. The energy and atomic number dependence is described in Sect. 2.1.

Coherent scattering is when bound atomic electrons scatter the incident photon, and the atom is neither excited nor ionized. The atomic cross section is approximately proportional to Z^2, and the energy dependence is $1/E^2$ above the energy where the atomic cross sections of coherent and incoherent scattering are the same; this energy is proportional to Z. At low photon energies, the energy dependence is approximately proportional to $1/E$ [4].

For soft tissues, the photoelectric effect dominates at energies lower than 20–30 keV, while the incoherent scattering dominates at higher energies. For water,

Å. C. Tedgren
Department of Health, Medicine and Caring Sciences, Linköping University, Linköping, Sweden

Center for Medical Image Science and Visualization (CMIV), Linköping University, Linköping, Sweden

Department of Medical Radiation Physics and Nuclear Medicine, Karolinska University Hospital, Stockholm, Sweden

A. Malusek
Department of Health, Medicine and Caring Sciences, Linköping University, Linköping, Sweden

Center for Medical Image Science and Visualization (CMIV), Linköping University, Linköping, Sweden coherent scattering contributes about ten percent to the total LAC at 20–40 keV. Its role is less important at other energies.

The energy-dependent mass attenuation coefficient, $\mu_m(E)$, defined as

$$\mu_m(E) = \mu(E)/\rho, \tag{2}$$

where ρ is the mass density of the material, reduces the strong dependence of the LAC on the material density. Values of mass attenuation coefficients can be obtained from databases like XCOM [5] or EPDL [6], which store evaluated cross section data, i.e., data obtained by combining theoretical calculations and experimental measurements.

The LAC of a mixture, μ, consisting of two base materials with mass attenuation coefficients $\mu_{m,1}$ and $\mu_{m,2}$ can be calculated as

$$\mu(E) = \rho\, w_{m,1}\, \mu_{m,1}(E) + \rho\, w_{m,2}\, \mu_{m,2}(E) = \rho_1\, \mu_{m,1}(E) + \rho_2\, \mu_{m,2}(E), \tag{3}$$

where ρ is the mass density of the mixture, and $w_{m,i}$ and ρ_i are the mass fraction and partial density, respectively, of base material $i = 1, 2$. The mass fractions are normalized so that $w_{m,1} + w_{m,2} = 1$. When the mixture is composed of other materials, mass fractions in Eq. (3) lose their physical meaning. Nevertheless, the LAC can still be decomposed as

$$\mu(E) = w_1\, \mu_1(E) + w_2\, \mu_2(E), \tag{4}$$

where the coefficients of proportionality w_1 and w_2 (called weight coefficients in this text) may become negative. In DECT, the two basis functions (or corresponding materials) are called the doublet. Note that for fixed w_1 and w_2 in Eq. (4), the basis functions predict the value of LAC at any energy E.

The AM method performs material decomposition according to Eq. (4) in the projection domain [1, 7]. The low and high tube voltage projections must be geometrically consistent. For this reason, AM does not work well with dual-source scanners. AM can be generalized to 3D helical geometry. Commonly used basis functions are the (approximate photoelectric effect, Compton scattering) doublet (PC) used, for instance, by the spectral detector computed tomography (CT) scanner IQon (Philips Healthcare), the (water, bone) doublet (WB) [7], and the (water, iodine) doublet (WI) [8] used by General Electric's DECT scanners. The PC doublet approximates the energy dependence of the photoelectric effect as

$$\mu_p(E) \sim E^{-3}. \tag{5}$$

However, some other authors also used $\mu_p(E) \sim E^{-2.8}$ [9], and Langeveld [10] proposed the $E^{-\beta(E)}$ dependence, where $\beta(E)$ is a third-order polynomial in $\ln(E/\text{keV})$, which combines contributions from the photoelectric effect and coherent scattering. The energy dependence of the Compton scattering is given by the Klein–Nishina cross section. The WB and WI doublets are derived from tabulated data of cross sections. The AM method uses the calculated weight coefficients w_1 and w_2 for the computation of monoenergetic images at any energy. This way, beam-hardening can be eliminated. A review of the performance and practical applications of the AM method can be found in Ref. [7].

DIRA performs the material decomposition in the image domain. Geometrical consistency between low and high tube voltage projections is not required. DIRA can be extended to 3D helical geometry [11]. Calculated forward projections together with the original projections are used to update the reconstructed image. The image is then decomposed to mass fractions of the base materials as well as the density. It is possible to use different base material doublets at different spatial positions in the image, for instance, for different organs. Alternatively, a three-material decomposition can be used. In this case, the mass density of the mixture is not a free parameter but is calculated from the assumption about the preservation of molar volumes.

Of interest is the minimum number of basis functions that can reliably represent the LAC of biological tissues in the medical CT energy range 20–150 keV. This problem is known as the intrinsic dimensionality of the cross section data. Williamson et al. [12] investigated biological tissues with effective atomic numbers $Z = 2, \ldots, 20$ in the energy range 20–1000 keV. For determination of individual (mass macroscopic) cross sections, $\mu_p(E)/\rho$ and $\mu_{inc}(E)/\rho$, where ρ is the mass density, they recommended a (water, polystyrene) doublet for $Z = 1, \ldots, 8$ and a (water, calcium chloride solution) doublet for $Z = 8, \ldots, 20$. On the other hand, they claimed that only one base material doublet is sufficient to represent LAC values. Bornefalk [13] applied the Principal Component Analysis (PCA) to LACs affected by uncertainties for $Z = 1, \ldots, 20$ in the energy range 25–120 keV and discovered an intrinsic dimensionality of 3 to 4. Alvarez claimed that two base materials are sufficient to represent the attenuation coefficients of biological tissues [14]. However, if a contrast agent with a high atomic number is present, then three or more base materials are needed.

The conflicting results on the dimensionality of the cross section data raise questions on the cause of this discrepancy and how the choice of the material base affects the material decomposition. This work aims to investigate the problem (i) in the ideal case of a direct decomposition of the LAC to basis functions and (ii) in the case when the material decomposition is performed by the AM and DIRA reconstruction algorithms on noiseless simulated projection data.

2 Theory

2.1 The Physics of Incoherent Scattering

The cross section for the scattering of a photon on a free electron was first derived by Klein and Nishina. It can be written as

$$\sigma_{KN}(\alpha) = 2\pi r_e^2 \frac{1+\alpha}{\alpha^2} \left[\frac{2(1+\alpha)}{1+2\alpha} - \frac{\ln(1+2\alpha)}{\alpha} \right] + \frac{\ln(1+2\alpha)}{2\alpha} - \frac{1+3\alpha}{(1+2\alpha)^2},$$
(6)

where $\alpha = E/m_ec^2 = E/510.975$ keV is the photon energy relative to the rest energy of electron, and r_e is the classical electron radius.

Tabulations of the atomic incoherent scattering cross section for bound (atomic) electrons combine the Klein–Nishina cross section for free electrons with the scattering function $S(x, Z)$, which takes into account the number of atomic electrons Z and the momentum transfer parameter x [15]. For large x, the function $S(x, Z)$ approaches Z.

2.2 The Physics of Polyenergetic Computed Tomography (CT) Measurements

In a CT scanner, an X-ray source emits X-rays that are attenuated by the object, e.g., a patient. Detectors register the X-rays after they have passed through the object.

The X-ray source emits photons in a range of energies. Two typical energy spectra for the tube voltages of 80 kV and 140 kV are illustrated in Fig. 1a. The X-ray beam intensity I at the detector position is given by

$$I = \int_0^{E_{\max}} D(E)\, N(E)\, \exp\left[-\int_{\mathcal{L}} \mu(x, y, E)\, dl \right] dE,$$
(7)

where $D(E)$ is the detector responsivity, $N(E)\, dE$ is the number of photons with energies in the interval $(E,\ E + dE)$, and $\mu(x, y, E)$ is the LAC of the object at position (x, y) for photons with energy E. It is noteworthy that $D(E) = E$ and $D(E) = 1$ for ideal energy-integrating and ideal photon-counting detectors, respectively. The inner integration in Eq. (7) is performed over a straight line $\mathcal{L}$ through the object. An unattenuated ray that does not pass the object gives

$$I_0 = \int_0^{E_{\max}} D(E)\, N(E)\, dE.$$
(8)

Projections P are then calculated by comparing Eqs. (7) and (8) according to

$$P = -\ln(I/I_0).$$
(9)

For a monoenergetic spectrum, the projections P_{mono} are equal to line integrals of the object function at energy E_{mono},

$$P_{\mathrm{mono}} = \int_{\mathcal{L}} \mu(x, y, E_{\mathrm{mono}})\, dl.$$
(10)

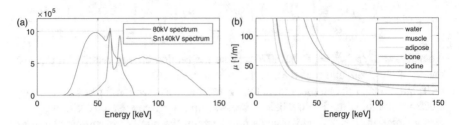

Fig. 1 (**a**) Energy spectrum of photons emitted from the X-ray tube at voltages of 80 kV (blue) and 140 kV (red). (**b**) Tabulated values of linear attenuation coefficients, $\mu(E)$, for water, adipose, muscle, compact bone, and iodine

This equation is also the Radon transform defined on the space of straight lines $\mathcal{L} \subset \mathbb{R}^2$. The reconstruction of $\mu(x, y, E_{mono})$ from projections P_{mono} in Eq. (10) constitutes an inverse problem. The most commonly used reconstruction algorithm is filtered backprojection (FBP); see, e.g., Ref. [16]. The LAC curve varies with the energy, and its shape differs for different materials. Figure 1b shows LAC curves for water, muscle, adipose tissue, and compact bone. Note that the curve for compact bone stands out slightly from the other curves regarding shape. It is also higher due to its larger density. The iodine curve stands out even more with its K-edge at 33.2 keV.

It is common to use the mathematically accurate reconstruction algorithm for monoenergetic projections also for polyenergetic projections P based on Eqs. (7), (8), and (9). Since this is an approximation of the Radon transform, there is a risk of artifacts. These are called beam-hardening artifacts, and they become more severe for highly attenuating objects. A common visual beam-hardening artifact is dark streaks between bones. Beam-hardening makes the attenuation values deviate from the true ones. It can be partially suppressed by conventional water beam-hardening correction; see, e.g., Ref. [16].

2.3 Alvarez and Macovski's Method

For human tissues, the linear attenuation coefficient $\mu(E)$ can be approximately decomposed as in Eq. (4), $\mu(E) = w_1 \mu_1(E) + w_2 \mu_2(E)$, where w_1 and w_2 are weight coefficients, and $\mu_1(E)$ and $\mu_2(E)$ are the LACs for two basis functions (doublet).

Alvarez and Macovski proposed a method that can perfectly exclude all beam-hardening artifacts, provided that Eq. (4) is valid. We can express weight coefficients depending on the position (x, y) within the material by modifying the Eq. (4) to

$$\mu(x, y, E) = w_1(x, y) \mu_1(E) + w_2(x, y) \mu_2(E). \tag{11}$$

Utilizing this in Eq. (7) gives

$$I = \int_0^{E_{max}} D(E) \, N(E) \, \exp\left[-\int_{\mathcal{L}} w_1(x, y)\, \mu_1(E) + w_2(x, y)\, \mu_2(E)\, dl \right] dE.$$
(12)

In the Alvarez and Macovski (AM) method, two different spectra are needed, one with lower energies and one with higher energies, such as the ones in Fig. 1a. Let $N_L(E)$ and $N_H(E)$ be the spectra for the low and high tube voltage, respectively. Let the normalized detector response be defined as

$$S_i(E) = D(E) \, N_i(E) \bigg/ \int_0^{E_{max}} D(E) \, N_i(E)\, dE,$$
(13)

where i stands for L or H. An insertion of Eqs. (12) and (8) into (9) gives the modeled dual-energy projections P_L and P_H:

$$P_L = -\ln\left[\int_0^{E_{max}} S_L(E) \exp\left(-\int_{\mathcal{L}} w_1(x, y)\mu_1(E)\, dl - \int_{\mathcal{L}} w_2(x, y)\mu_2(E)\, dl \right) dE \right],$$

$$P_H = -\ln\left[\int_0^{E_{max}} S_H(E) \exp\left(-\int_{\mathcal{L}} w_1(x, y)\mu_1(E)\, dl - \int_{\mathcal{L}} w_2(x, y)\mu_2(E)\, dl \right) dE \right].$$
(14)

The projections of the weight coefficients w_1 and w_2 can be named P_{w_1} and P_{w_2}, respectively, i.e.,

$$P_{w_i} = \int_{\mathcal{L}} w_i(x, y)\, dl,$$
(15)

where i stand for 1 or 2. Insertion in Eq. (14) gives

$$P_L = -\ln\left[\int_0^{E_{max}} S_L(E) \exp\left(-P_{w_1} \mu_1(E) - P_{w_2} \mu_2(E) \right) dE \right],$$

$$P_H = -\ln\left[\int_0^{E_{max}} S_H(E) \exp\left(-P_{w_1} \mu_1(E) - P_{w_2} \mu_2(E) \right) dE \right].$$
(16)

2.3.1 Solution of the Equation System

The next step is to find P_{w_1} and P_{w_2} so that

$$P_L \left(P_{w_1}, P_{w_2} \right) - P_{M,L} = 0,$$

$$P_H \left(P_{w_1}, P_{w_2} \right) - P_{M,H} = 0,$$
(17)

where $(P_{M,L}, P_{M,H})$ are the *measured* dual-energy projection values. Several solutions have been suggested in the literature. According to Ying et al. [17], the solution to the equation system (17) can be found numerically via the two-dimensional Newton–Raphson method [18], which uses the Jacobian matrix of the transformation defined by Eqs. (16).

The method may result in either P_{w1} or P_{w2} negative as commonly observed for the WB or WI doublets. The resulting negative weight coefficients do not break the AM method, nevertheless, they complicate physical interpretation. Non-negative values may be required for the PC doublet, see the discussion in Sect. 5. In this case, Ying et al. [17], who considered the PC doublet only, used the following method. If the point (P_{w_1}, P_{w_1}) falls outside the first quadrant, then the task is a minimization problem with constraints $P_{w_1} \geq 0$ and $P_{w_2} \geq 0$. The solution is a point (P_{w_1}, P_{w_2}) corresponding to the minimum

$$\min_{(P_{w_1}, P_{w_2})} \left(\left[P_L \left(P_{w_1}, P_{w_2} \right) - P_{M,L} \right]^2 + \left[P_H \left(P_{w_1}, P_{w_2} \right) - P_{M,H} \right]^2 \right), \qquad (18)$$

which can be found iteratively via the Newton–Gauss method.

2.3.2   Reconstruction of Base Material Images and Monoenergetic Images

Base material images of the weight coefficients $w_1(x, y)$ and $w_2(x, y)$ can then be reconstructed from P_{w1} and P_{w2} by using, e.g., FBP. Moreover, virtual monoenergetic images $\mu(x, y, E_{mono})$ at any desired energy can be computed by inserting $w_1(x, y)$, $w_2(x, y)$, and $E = E_{mono}$ in Eq. (11).

2.4   Dual-energy Iterative Reconstruction Algorithm (DIRA)

DIRA is a model-based iterative reconstruction algorithm that can perform two- and three-material decomposition of the imaged object to any number of doublets or triplets [2]. Calculated forward projections together with the measured projections are used to update the reconstructed image iteratively. For each iteration, the mathematical decomposition of the image to base materials is also improved. This section describes a configuration using the WB and WI doublets simultaneously; no triplets are used.

2.4.1   Computation of Forward Polyenergetic Projections in DIRA

By rewriting Eq. (3) slightly, the LAC can be expressed as

$$\mu(x, y, E) = \rho(x, y) \sum_{i=1}^{2} w_{m,i}(x, y) \, \mu_{m,i}(E).$$
(19)

The X-ray beam intensity I at the detector position is given by Eq. (7). Inserting Eq. (19) in this equation gives

$$I = \int_0^{E_{max}} D(E) N(E) \exp\left[-\int_{\mathcal{L}} \left(\rho(x, y) \sum_{i=1}^{2} w_{m,i}(x, y) \, \mu_{m,i}(E) \right) dl \right] dE$$
(20)

$$= \int_0^{E_{max}} D(E) N(E) \exp\left[-\sum_{i=1}^{2} \mu_{m,i}(E) \underbrace{\int_{\mathcal{L}} w_{m,i}(x, y) \, \rho(x, y) \, dl}_{l_i} \right] dE.$$
(21)

The line integrals l_1 and l_2 are calculated with Joseph's method [19]. Equation (21) shows that the calculation of line integrals can be performed on the weighted mass densities of the two base materials. The influence of the X-ray spectrum and mass attenuation coefficients can be taken into account afterwards. Direct computation of Eq. (21) is considerably faster than computation of Eq. (20). This is an important observation because it significantly reduces the calculation time for DIRA.

The detector-weighted spectral energy $D(E) N(E)$, and the mass attenuation coefficients as functions of energy $\mu_{m,i}(E)$, can be re-sampled to a common grid E_k, $k = 0, .., K$. Then Eq. (21) can be computed using the Simpson's formula,

$$I \approx \frac{1}{2} \sum_{k=1}^{K-1} D(E_k) N(E_k) \exp\left[-\sum_{i=1}^{2} \mu_{m,i}(E_k) l_i \right] (E_{k+1} - E_{k-1}).$$
(22)

Finally, the polyenergetic projections P_L and P_H are received by inserting I into Eq. (9).

2.4.2 Choosing the Two Effective Energies in DIRA

Two effective energies $E_{eff,L}$ and $E_{eff,H}$ are chosen in DIRA by using the following method. The effective attenuation coefficient μ_{eff} for water is calculated by using the weighted mean of the X-ray spectrum and the attenuation curve $\mu(E)$ for water,

$$\mu_{eff} = \frac{\int_0^{E_{max}} D(E) N(E) \mu(E) \, dE}{\int_0^{E_{max}} D(E) N(E) \, dE}.$$
(23)

Then, using the $\mu(E)$-curve for water, the energy value corresponding to μ_{eff} is taken as the effective energy. The effective energies $E_{eff,L}$ and $E_{eff,H}$ are obtained by setting $N(E) = N_L(E)$ and $N(E) = N_H(E)$, respectively, in Eq. (23). As before, L

stands for low and H stands for high. It is not necessary to use exactly these values, but experimentally we have found that they cause DIRA to converge faster.

2.4.3   Computation of Forward Monoenergetic Projections in DIRA

Monoenergetic projections, for the $E_{\mathrm{eff,L}}$ and $E_{\mathrm{eff,H}}$ energies, are calculated as

$$P_{\mathrm{mono,L}} = \mu_{\mathrm{m},1}(E_{\mathrm{eff,L}})l_1 + \mu_{\mathrm{m},2}(E_{\mathrm{eff,L}})l_2,$$
$$P_{\mathrm{mono,H}} = \mu_{\mathrm{m},1}(E_{\mathrm{eff,H}})l_1 + \mu_{\mathrm{m},2}(E_{\mathrm{eff,H}})l_2. \tag{24}$$

2.4.4   Base Material Decomposition in DIRA

The base material decomposition is made possible by forming an equation system with Eq. (19) for the two effective energies

$$\hat{\mu}_{\mathrm{m}}(x, y, E_{\mathrm{eff,L}}) = w_{\mathrm{m},1}(x, y)\rho(x, y)\mu_{\mathrm{m},1}(E_{\mathrm{eff,L}}) + (1-w_{\mathrm{m},1}(x, y))\mu_{\mathrm{m},2}(E_{\mathrm{eff,L}}),$$
$$\hat{\mu}_{\mathrm{m}}(x, y, E_{\mathrm{eff,H}}) = w_{\mathrm{m},1}(x, y)\rho(x, y)\mu_{\mathrm{m},1}(E_{\mathrm{eff,H}}) + (1-w_{\mathrm{m},1}(x, y))\mu_{\mathrm{m},2}(E_{\mathrm{eff,H}}), \tag{25}$$

where $w_{\mathrm{m},1} + w_{\mathrm{m},2} = 1$. The $\hat{\mu}_{\mathrm{m}}$ functions are the current LAC functions. They are updated with the iterations. From (25), $w_{\mathrm{m},1}(x, y)$ and $\rho(x, y)$ are solved.

2.4.5   Summary of DIRA

The complete DIRA algorithm is now illustrated in Fig. 2 and described below.

1. Measured projections $P_{\mathrm{M,L}}$ and $P_{\mathrm{M,H}}$ are obtained for low- and high-energy spectra, respectively. They are reconstructed with filtered backprojection (FBP) to corresponding μ_{L} and μ_{H}. The initial reconstruction (iteration 0) is preceded by a conventional water beam-hardening correction.
2. For the phantoms considered in this work, a threshold segmentation is performed to separate regions containing iodine from those containing other materials. The results is stored in two images $\mu_{\mathrm{T,L}}$ and two images $\mu_{\mathrm{T,H}}$.
3. A base material decomposition gives density and base material mass fraction images μ_B.
4. Monoenergetic forward projections, $P_{\mathrm{mono,L}}$ and $P_{\mathrm{mono,H}}$, are generated for the low and high effective energies, $E_{\mathrm{eff,L}}$ and $E_{\mathrm{eff,H}}$, respectively. They are then reconstructed to $\mu_{\mathrm{mono,L}}$ and $\mu_{\mathrm{mono,H}}$.
5. Polyenergetic forward projections P_{L} and P_{H} are generated and compared with the simulated measured projections. Reconstruction by FBP gives updates $\Delta\mu_{\mathrm{L}}$

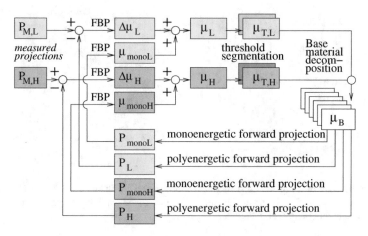

Fig. 2 Flowchart of the dual-energy iterative reconstruction algorithm DIRA

and $\Delta\mu_H$. The addition to $\mu_{\mathrm{mono,L}}$ and $\mu_{\mathrm{mono,H}}$ gives μ_L and μ_H for the next iteration.

3 Methods

3.1 Visual Investigation of the Dimensionality of Relevant Materials

Material decomposition to the WB doublet was performed for elements with $Z = 1, \ldots, 20$, Ti, Zn, I, Ba, Ce, Au, and Gd and materials like water, lipid, protein, adipose, muscle, compact bone, and femora spongiosa. Reasons for choosing these materials were as follows. Most of the elements comprising human tissues have $Z \leq 20$. Ti may be used for implants, Zn can be found in small amounts in the prostate, and I, Ba, Ce, and Gd can be used as contrast agents. Au is used as fiducial markers and nanoparticles. Soft tissues consist mainly of water, lipids, and proteins. Bones consist of compact bone and spongiosa. True mass attenuation coefficients, $\mu_{\mathrm{m,tab}}$, for these materials were either taken directly from the EPDL97 library [20] or derived from elemental compositions taken from Ref. [2]. Corresponding true linear attenuation coefficients were obtained as $\mu_{\mathrm{tab}} = \rho\mu_{\mathrm{m,tab}}$, where ρ is the mass density of the material. The weight coefficients w_1 and w_2 in Eq. (4) are linearly proportional to the density of the decomposed material. Consequently, the ratio μ/μ_{tab}, which was used to assess the approximation quality visually, does not depend on the density ρ. To evaluate the ratio, the density was set to $\rho = 1\,\mathrm{g\,cm^{-3}}$ and the weight coefficients w_1 and w_2 were obtained by solving an equation system consisting of Eq. (4) at the energies of 50 and 88 keV.

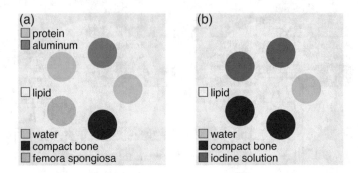

Fig. 3 Schematic drawing of the phantom without (**a**) and with (**b**) the iodine-water solution inserts. The lipid cylinders with the diameter of 316 mm contained 5 rod inserts with the diameter of 70 mm containing different materials

Table 1 Elemental composition and corresponding density for phantom and base materials. The values are rounded

Material	Mass fraction (%)	Density (g/cm^3)
Lipid	H 11.8, C 77.3, O 10.9	0.92
Protein	H 6.6, C 53.4, N 17.0, O 22.0, S 1.0	1.35
Water	H 11.2, O 88.8	1.00
Compact bone	H 3.6, C 15.9, N 4.2, O 44.8, Na 0.3, Mg 0.2, P 9.4, S 0.3, Ca 21.3	1.92
Femora marrow	H 9.4, C 38.5, N 2.2, O 43, Na 0.2, P 2.2, S 0.3, Cl 0.1, Ca 4.1	1.12
Iodine solution	I 6.0, H 10.5, O 83.5	1.05

3.2   X-ray Spectra and Phantoms

Photon energy spectra for X-ray tube voltages of 80 and 140 kV were used; see Fig. 1a. The latter spectrum was filtered with an additional tin filter.

Mathematical models of two cylindrical phantoms filled with lipid and containing five rod inserts were used, see Fig. 3. Rod inserts of the first phantom consisted of water, protein, compact bone, femora spongiosa, and aluminum. Rod inserts of the second phantom consisted of water, compact bone, and iodine-water solution. The material compositions are shown in Table 1.

3.3   Setup for the AM Algorithm

The AM algorithm described in Sect. 2.3 was implemented according to Ying et al. [17]. For the WB and WI doublets, the Newton–Raphson method was used, and negative solutions were allowed. For the PC doublet, the Newton–Raphson method was modified to eliminate negative solutions by setting negative P_{w1} or P_{w2} to 0 in every iteration. The minimization described by expression (18) and performed by the Newton–Gauss method was not used. The Newton–Raphson algorithm was set to stop when the increment was small enough or the number of iterations reached the maximum value of 10.

Projections of the phantom were calculated with Drasim [21] in a fan beam geometry described in Ref. [2] and rebinned to parallel projections. The *measured* projections $P_{M,L}$ and $P_{M,H}$ were simulated using 80 and 140 kV spectra, respectively, see Fig. 1a. An ideal energy-integrating detector with the response $S(E) = E\,N(E)$ was assumed.

AM with the PC, WB, and WI doublets was used for the phantoms with and without the iodine insert. The AM reconstruction resulted in two images showing the base material weight coefficients w_1 and w_2. Averages of w_1 and w_2 were taken in regions of interest (ROIs) inside the rod inserts and the water cylinder. From these values, LACs as functions of energy in the range 20–150 keV were obtained using Eq. (4). Also, three monoenergetic images for 30, 50, and 88 keV were calculated.

3.4 Setup for DIRA

DIRA was implemented according to the description in Sect. 2.4. Projections of the phantom were calculated the same way as for the AM method. Only the phantom with iodine was processed.

Low and high effective energies were $E_{\mathrm{eff,L}} = 50\,\mathrm{keV}$ and $E_{\mathrm{eff,H}} = 88\,\mathrm{keV}$, respectively. Monoenergetic forward projections $P_{\mathrm{mono,L}}$ and $P_{\mathrm{mono,H}}$ at these energies were generated.

The threshold segmentation to $\mu_{\mathrm{T,L}}$ and $\mu_{\mathrm{T,H}}$ was used to separate regions with iodine from regions with other materials. To separate iodine from soft tissue, a threshold at 30 m^{-1} for μ_{L} was used. To distinguish between iodine and bone, $k = \mu_{\mathrm{L}}/\mu_{\mathrm{H}}$ was calculated. If $k>2.4$, iodine solution was supposed. According to Table 2, compact bone has $k = 79.2/38.9 = 2.04$ and iodine solution has $k = 100.0/34.7 = 2.88$, which justifies the choice of a threshold at 2.4.

The base material decomposition were performed as follows. In the iodine regions, two-material decomposition to the WI doublet was used. In the bone and soft tissue regions, the WB doublet was used.

The loop was iterated 16 times, but the algorithm converged earlier. The final μ_{L} and μ_{H} were the reconstructions at 50 and 88 keV, respectively. A virtual monoenergetic image at 30 keV was obtained by inserting the density and base material mass fraction images (μ_B) along with corresponding mass attenuation coefficients in Eq. (19).

Table 2 LACs (in m^{-1}) reconstructed by AM using the PC, WB, and WI doublets, and DIRA using both WB and WI doublets for the phantom with iodine. True values are also listed

Energy	Material	True	AM with PC	AM with WB	AM with WI	DIRA
30 keV	Water	37.6	37.4	37.1	37.6	37.6
	Compact bone	244.9	247.9	245.9	91.0^a	244.4
	Iodine solution	90.3	414.6^a	412.1^a	90.5	89.4
50 keV	Water	22.7	22.6	22.6	22.8	22.7
	Compact bone	79.2	79.6	79.4	77.3^b	79.4
	Iodine solution	100.0	105.6^b	105.5^b	100.3	100.3

^{a}Values deviating more than 62%
^{b}Values deviating more than 2.4%

4   Results

The ability of the WB doublet to represent the material of interest is represented by the $\mu(E)/\mu_{tab}(E)$ ratio. This ratio, obtained by direct application of material decomposition via Eq. (4) at the energies of 50 and 88 keV, is plotted in Figs. 4 and 5 for elements with $Z = 1, \ldots, 20$ and selected common human tissues, respectively.

Note that the elements are well represented for energies above ≈35 keV. For energies in the range 20–35 keV, the discrepancy is larger, especially for H, He, Li, Be, B. However, these substances, except for H, are typically not found in the human body. Cross sections of compounds containing H are typically dominated by the other elements. In Fig. 5, the $\mu(E)/\mu_{tab}(E)$ ratios for water, bone, lipid, protein, adipose, muscle, and femora spongiosa are plotted; the discrepancy in the range 20–35 keV was small.

The ratios for Ti, Zn, I, Ba, Ce, Au, and Gd are plotted in Fig. 6. The ability of the WB doublet to represent these elements decreases with increasing atomic number of the element. It is still good for Ti with the atomic number of 22, becomes worse for Zn with atomic number 30, and very inaccurate for Gd and Au with atomic numbers of 64 and 79, respectively. Large discrepancy can be observed for I, Ba, and Ce (atomic numbers of 53, 56, and 58) at energies below the K-edges of these elements (30–40 keV).

4.1   Analysis of Reconstructed Data for the Phantom Without an Iodine Insert

Figure 7 shows images reconstructed by AM using the WB doublet at 30, 50, and 88 keV. There was a good agreement with tabulated values. The largest difference was for the aluminum rod at 30 keV, with the value 299.6 m^{-1} measured in a ROI and the tabulated 303.8 m^{-1}, giving a ratio of 299.6/303.8 ≈ 0.986. This value agrees with aluminum in Fig. 8a. Figure 8 shows the $\mu(E)/\mu_{tab}(E)$ ratio obtained by AM using the WB, PC, and WI doublets and the phantom without iodine. The

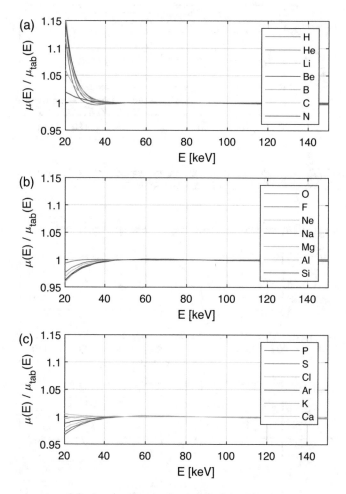

Fig. 4 The LAC predicted by the (water, bone) doublet relative to the tabulated LAC, $\mu(E)/\mu_{tab}(E)$, as a function of energy for elements with $Z = 1, \ldots, 20$

WB and PC doublets predicted the LAC well for $E > 40$ keV. In the range 20–35 keV, the relative difference between the LACs was larger, but it was still less than 5%. In the case of WI, the K edge of iodine at 33.2 keV caused large discrepancies between the LACs in the range 20–40 keV. These discrepancies lead to a notable beam-hardening artifact between the compact bone and aluminum inserts at 50 keV, see also Fig. 9.

Reconstructed images of the phantom without iodine are shown in Fig. 9 using a reduced range of LAC values. Contrary to the WB and PC doublets, the WI doublet produced a clearly visible beam-hardening artifact. Results from experiments with DIRA using the WB doublet (not shown here) were similar to those with AM using WB.

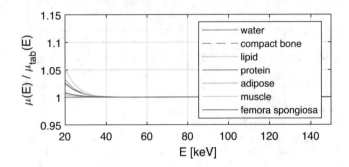

Fig. 5 The LAC predicted by the (water, bone) doublet relative to the tabulated LAC, $\mu(E)/\mu_{\mathrm{tab}}(E)$, as a function of energy for selected materials

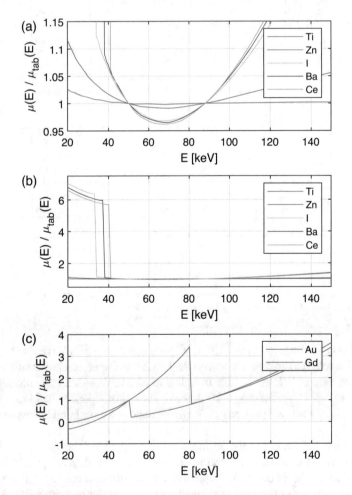

Fig. 6 The LAC predicted by the (water, bone) doublet relative to the tabulated LAC, $\mu(E)/\mu_{\mathrm{tab}}(E)$, as a function of energy for Ti, Zn, I, Ba, Ce, Au, and Gd. The first five elements are plotted using the fine (**a**) and coarse (**b**) ranges on the y-axis. The last two elements are plotted using the coarse (**c**) range on the y-axis

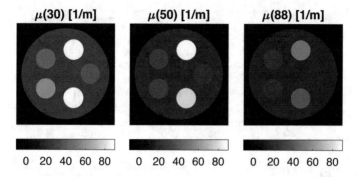

Fig. 7 LACs (in m^{-1}) for the phantom without iodine reconstructed by AM at 30, 50, and 88 keV with the WB doublet

The PC doublet did not provide the correct fractions of the photoelectric and Compton LACs. Nevertheless, Fig. 10 shows that it agreed with the tabulated (photo + incoherent + coherent) LAC when both components were added together. This shows that the PC doublet with the E^{-3} dependence can be used in the AM for beam-hardening removal but not for the determination of true photoelectric and Compton fractions.

4.2 Analysis of Reconstructed Data for the Phantom with an Iodine Insert

Images of the phantom with an iodine insert were reconstructed at 30, 50, and 88 keV by (i) AM using the PC, WB, and WI doublets, and (ii) DIRA using the WB and WI doublets simultaneously. The images are shown in Fig. 11 using a reduced range of LAC values. The image for AM using the PC doublet is not shown since it was indistinguishable from the image for AM using the WB doublet. The AM method produced clearly visible beam-hardening artifacts for all three doublets. DIRA reconstructed the phantom without such artifacts. LAC values for 30 and 50 keV are given in Table 2, where deviating values are highlighted. For 88 keV, the discrepancies were small for all methods; the true values were 17.7, 38.9, and 34.7 m^{-1} for water, bone, and iodine solution, respectively.

5 Discussion

The largest discrepancies in the prediction of the LAC are expected at low energies when a combination of both low- and high-Z materials is used. This was observed for (i) the WB doublet predictions for elements with $Z = 1, \ldots, 5$ (Fig. 4) in the 20–

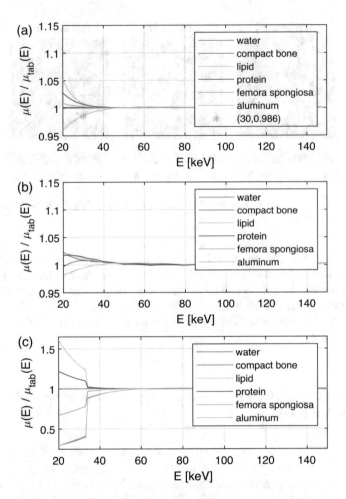

Fig. 8 The predicted LAC relative to the tabulated LAC, $\mu(E)/\mu_{tab}(E)$, as a function of energy calculated by AM using the (water, bone) (**a**), (approximate photoelectric effect, Compton scattering) (**b**), and (water, iodine) (**c**) doublets for the phantom without iodine. Note the larger range on the y-axis in panel (**c**)

25 keV region, (ii) the WB doublet approximating high Z materials like Zn in the energy range 20–30 keV (Fig. 6), (iii) high Z contrast agents like I, Ba, and Ce below the K-edge energy (Fig. 6), and (iv) the WI doublet predictions for the materials in the phantom without iodine (Fig. 8). Of special interest is that the deviation was very large when contrast agents were involved, either as a phantom material or as part of the WI doublet; see Figs. 6 and 8c. This observation is in line with Alvarez's statement that two base materials are not sufficient in this case [14]. The ability of the WB doublet to represent high Z elements such as Au and Gd was poor for the entire investigated energy range 20–140 keV (Fig. 6c).

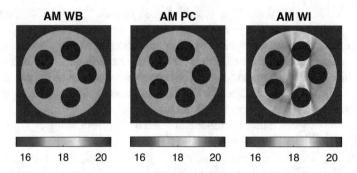

Fig. 9 LACs (in m^{-1}) reconstructed by AM at 50 keV for the WB, PC, and WI doublets and the phantom without iodine

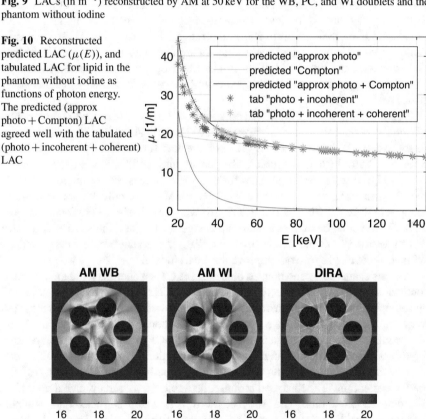

Fig. 10 Reconstructed predicted LAC ($\mu(E)$), and tabulated LAC for lipid in the phantom without iodine as functions of photon energy. The predicted (approx photo + Compton) LAC agreed well with the tabulated (photo + incoherent + coherent) LAC

Fig. 11 LACs (in m^{-1}) for the phantom with iodine reconstructed at 50 keV by AM using the WB and WI doublets and by DIRA

The material decomposition in the AM method to the PC doublet neglects the coherent scattering contribution, but, on the other hand, it uses the Klein–Nishina cross section, which overestimates the incoherent scattering contribution at low energies, and thus, to a certain degree, compensates for the neglected coherent scattering. This approximation can lead to inaccurate fractions of the photoelectric effect and Compton scattering components. Nevertheless, the sum of both contributions can be biased much less. In our experiments, this behavior can be seen in Fig. 10. A compensation via the energy dependence of the photoelectric effect as $\mu_p(E) \sim E^{-2.8}$ suggested by Weaver and Huddleston [9] gave much worse results than $\mu_p(E) \sim E^{-3}$ (results are not presented here).

Figure 4 shows that it may be difficult to reach the dimensionality of 3–4 stated by Bornefalk [13] at the energy range of 35–150 keV for elements with $Z = 1, \ldots, 20$. In this energy range, the dimensionality of 2 is more likely, i.e., one doublet can predict the LAC values of the elements. All the major differences between the LACs of elements are in the energy range 20–35 keV. The relative numbers of photons in the low-energy part of the low- and high-energy X-ray spectra (Fig. 1) are small (<7.9% and <0.15%, respectively) in the range 20–35 keV (Fig. 1). Moreover, in clinical applications, most of these photons are absorbed by adult patient bodies. The situation may be different for children and Spectral CT, where low-energy photons may pass the small bodies and contribute to the low-energy channel at the 20–40 keV.

Figure 4 showed how well elements with $Z = 1, \ldots 20$ can be approximated with the WB doublet in the energy range 20–150 keV. A similar investigation was performed by Williamson et al. [12] (Figure 2a) for the (polystyrene, calcium chloride solution) (PCCS) doublet. The performance of PCCS was better than our WB (in Fig. 4) in the lower energies for many Z elements. For example, at 20 keV and nitrogen, the relative difference was 2% for WB, but only 1% for PCCS. For energies >40 keV, both WB and PCCS gave very small relative differences, below 0.1%. On the other hand, for calcium and energies >110 keV, the relative difference was ≈2% for PCCS but below 0.1% for WB. The double doublet suggested by Williamson et al. [12] did not improve the result for Ca.

Photons scattered by the imaged object or the CT scanner may add an unwanted component to the signal registered by the detector. This component can be estimated by computations or measurements using additional detector elements and subtracted from the measured signal. If a correction for scattered radiation is not used, the reconstruction algorithm reduces the values of LACs in the image to compensate for the additional signal. Model-based reconstruction algorithms like DIRA can include the simulation of the scatter into the iterative loop. The quality of the scatter correction is then a compromise between the accuracy and duration of the computer simulation. The AM method expects scatter-corrected data as input. Current medical CT scanners typically perform scatter-correction. This work did not simulate scattered radiation; we expected its effect on images reconstructed from scatter-corrected projections would be low.

In Fig. 10, the basis function for the photoelectric effect in the PC doublet dominates for energies below 22 keV. In comparison, the basis function for Compton scattering dominates for energies above 22 keV. This separation of basis functions also results in positive weight coefficients for the PC doublet. On the other hand, the basis functions for water and bone in the WB doublet have similar shapes (Fig. 1b);

the only main difference is that the function for bone is steeper at low energies than that of water due to the photoelectric effect on calcium. A difference between the two basis functions must be calculated to represent high-Z materials with even steeper functions at low energies; hence one of the weight coefficients must be negative.

Projection data are affected by quantum and electronic noise. This noise is propagated through the image reconstruction and material decomposition algorithms and affects the resulting weight coefficients. Experiments (not shown here) with quantum noise added to the projection data indicated that a quantum noise level common in medical diagnostic imaging did not affect the stability of the AM method. There were no visual differences between monoenergetic images at 70 keV reconstructed by the PC and WB doublet. The stability of DIRA with respect to noise was demonstrated in Ref. [2].

The presented work used computer simulations to eliminate machine and quantum noise-related artifacts. In practical applications, however, these will also affect the results, especially for low tube loads, and the quantum noise will decrease the precision of the base material weights. The potentially different sensitivities to the noise of AM and DIRA and the methods for noise reduction in both algorithms are subjects for future research.

6 Conclusion

The PC and WB doublets accurately predicted the LAC values for human tissues and elements with $Z = 1, \ldots, 20$, in the 20–150 keV range, though there was a small (<5%) discrepancy in the 20–35 keV range. The WI doublet did not represent the tissues as well as PC and WB; the largest discrepancies (>50% in some cases) were in the 20–40 keV range.

LACs reconstructed with the AM and DIRA followed this trend. AM produced artifacts when iodine was present in the phantom together with human tissues since AM can only work with one doublet. It was shown that these artifacts could be avoided with DIRA using the WB doublet for bone and soft tissues and the WI doublet for the iodine solution.

Funding
This work was supported by Cancerfonden [CAN 2017/1029, CAN 2018/622]; ALF Grants Region Östergötland [LiO-602731]; Patientsäkerhetsforskning Region Östergötland [LiO-724181]; and Vetenskapsrådet [VR-NT 2016-05033].

Conflict of Interest

The authors declare no conflicts of interest with regards to this work.

Acknowledgments David Ballestero is acknowledged for implementing AM.

References

1. Alvarez RE, Macovski A. Energy-selective reconstructions in X-ray computerised tomography. Phys Med Biol. 1976;21:733–44. ISSN: 00319155. https://doi.org/10.1088/0031-9155/21/5/002. http://iopscience.iop.org/0031-9155/21/5/002
2. Malusek A, Magnusson M, Sandborg M, Carlsson GA. A model-based iterative reconstruction algorithm DIRA using patient-specific tissue classification via DECT for improved quantitative CT in dose planning. Med Phys. 2017;44(6):2345–2357. https://doi.org/10.1002/mp.12238. http://onlinelibrary.wiley.com/doi/10.1002/mp.12238/full
3. Podgoršak EB. Radiation physics for medical physicists. Cham: Springer International Publishing; 2016. OCLC: 1224247725. ISBN: 978-3-319-25382-4. https://doi.org/10.1007/978-3-319-25382-4
4. Carlsson GA, Dance D. Interactions of photons with matter. In: Mayles P, Nahum A, Rosenwald J-C. editors. Handbook of radiotherapy physics: theory and practice. Milton Park: Taylor & Francis; 2007, p. 57–74. OCLC: 7350507061
5. Berger MJ, Hubbel JH, Seltzer SM, Chang J, Coursey JS, Sukumar R, Zucker DS, Olsen K. XCOM: photon cross sections database. NIST. Last Update: November 2010; Sept. 17, 2009. https://www.nist.gov/pml/xcom-photoncross-sections-database (visited on 06/14/2022)
6. Cullen DE. EPICS2017: april 2019 status report. IAEA-NDS-228. Vienna: International Atomic Energy Agency; 2019. p. 22. https://www-nds.iaea.org/publications/iaea-nds/iaea-nds-0228.pdf
7. Heismann BJ, Schmidt BT, Flohr T. Spectral computed tomography. Bellingham: SPIE Press; 2012. ISBN: 9780819492579. https://doi.org/10.1117/3.977546
8. Mendonca PRS, Lamb P, Sahani DV. A flexible method for multi-material decomposition of dual-energy CT images. IEEE Trans Med Imag. 2014;33(1):99–116. ISSN: 0278-0062. https://doi.org/10.1109/TMI.2013.2281719. http://ieeexplore.ieee.org/lpdocs/epic03/wrapper.htm?arnumber=6600785
9. Weaver JB, Huddleston AL. Attenuation coefficients of body tissues using principal-components analysis. Med Phys. 1985;12(1):40–5. ISSN: 0094-2405. https://doi.org/10.1118/1.595759. http://scitation.aip.org/content/aapm/journal/medphys/12/1/10.1118/1.595759
10. Langeveld WGJ. Effective atomic number, mass attenuation coefficient parameterization, and implications for high-energy X-ray cargo inspection systems. Phys Procedia. 2017;90:291–304. ISSN: 18753892. https://doi.org/10.1016/j.phpro.2017.09.014. https://linkinghub.elsevier.com/retrieve/pii/S1875389217301736 (visited on 06/09/2022)
11. Magnusson M, Björnfot M, Tedgren ÅC, Carlsson GA, Sandborg M, Malusek A. DIRA-3D—a model-based iterative algorithm for accurate dual-energy dual-source 3D helical CT. Biomed Phys Eng Exp. 2019;5(6):065005. ISSN: 2057-1976. https://doi.org/10.1088/2057-1976/ab42ee. https://iopscience.iop.org/article/10.1088/2057-1976/ab42ee
12. Williamson JF, Li S, Devic S, Whiting BR, Lerma FA. On two-parameter models of photon cross sections: application to dual-energy CT imaging: two-parameter cross-section models. Med Phys. 2006;33(11):4115–29. ISSN: 00942405. https://doi.org/10.1118/1.2349688. http://doi.wiley.com/10.1118/1.2349688

13. Bornefalk H. XCOM intrinsic dimensionality for low-Z elements at diagnostic energies. Med Phys. 2012;39(2):654. ISSN: 00942405. https://doi.org/10.1118/1.3675399. http://online.medphys.org/resource/1/mphya6/v39/i2/p654_s1

14. Alvarez RB. Dimensionality and noise in energy selective x-ray imaging. Med Phys. 2013;40(11):111909. ISSN: 00942405. https://doi.org/10.1118/1.4824057. http://doi.wiley.com/10.1118/1.4824057

15. Hubbell JH. Summary of existing information on the incoherent scattering of photons, particularly on the validity of the use of the incoherent scattering function. Radiat Phys Chem. 1997;50(1):113–24

16. Kak AC, Slaney M. Principles of computerized tomographic imaging. Philadelphia: Society for Industrial and Applied Mathematics; 2001. ISBN: 978-0-89871-494-4. OCLC: 46320986

17. Ying Z, Naidu R, Crawford CR. Dual energy computed tomography for explosive detection. J Xray Sci Technol 2006;14(4):235–256. http://iospress.metapress.com/index/D6KW951WJ5RFPR5U.pdf

18. Bertsekas DP. Nonlinear programming. Belmont: Athena Scientific; 1999. ISBN: 978-1-886529-00-7. OCLC: 1131554643

19. Joseph PM. An improved algorithm for reprojecting rays through pixel images. IEEE Trans Med Imaging. 1982;1(3):192–196. http://ieeexplore.ieee.org/xpls/abs_all.jsp?arnumber=4307572

20. Cullen DE, Hubbell JH, Kissel L. EPDL97: the evaluated photon data library, '97 version, 1997. Livermore: University of California, Lawrence Livermore National Laboratory; 1997. UCRL-50400

21. Stierstorfer K. DRASIM: A CT-simulation tool. Internal report. Siemens Medical Engineering; 2007.

Spectral Imaging in Photon-Counting CT with Data Acquired in Interleaved/Gapped Spectral Channels

Xiangyang Tang, Yan Ren, Huiqiao Xie, and Arthur E. Stillman

1 Introduction

Enormous effort has been invested in the research and development of spectral CT implemented via x-ray detection based on both energy integration [1, 2] and photon counting [3, 4] to meet the escalating challenges imposed by the advancements in clinical applications of diagnostic imaging. Compared to the energy-integration-based dual-energy CT (DECT), photon-counting spectral CT has a few fundamental advantages in its data acquisition and image formation to fulfill the requirements set by spectral imaging [3–5]. It is now exciting for us to witness that state-of-the-art photon-counting spectral CT has been cleared by the regulatory bodies for clinical applications [6], while the conventional energy-integration-based DECT is still adding significant value to the management of cardiovascular, oncologic, and neurovascular diseases.

For energy-integration spectral CT, e.g., dual-energy CT (DECT), there are three ways for spectral channelization in data acquisition [3]. The first technique is approached by switching, either fast or slow (sequential scan), the peak voltage of x-ray source during data acquisition [1]. Of course, the fast-switching outperforms the slow one in situations wherein the tissues/organs to be imaged are in motion, such as cardiovascular, respiratory, or other involuntary visceral motion [1]. The second one is carried out using a layered detector, wherein the spectral channels are formed by making use of the natural phenomenon that the x-ray photon at higher energy propagates a larger distance (depth) than those at lower energy [7]. The DECTs implemented in these two ways have been extending CT's clinical utility, although they have not made full use of the available spectral information (and thus reach

X. Tang (✉) · Y. Ren · H. Xie · A. E. Stillman
Department of Radiology and Imaging Sciences, Emory University School of Medicine, Atlanta, GA, USA
e-mail: xiangyang.tang@emory.edu

© The Author(s), under exclusive license to Springer Nature Switzerland AG 2023
S. Hsieh, K. (Kris) Iniewski (eds.), *Photon Counting Computed Tomography*,
https://doi.org/10.1007/978-3-031-26062-9_9

the achievable imaging performance). As has been reported in the literature, inter-channel spectral overlapping inevitably exists in those approaches, which may, to various extent, decrease what can be achieved by DECT in signal detection and noise suppression [8–10].

The third existing art of spectral channelization in energy-integration-based spectral CT is the dual-source-dual-detector (DSDD) technique, wherein each of the x-ray sources works at distinct (low or high) peak voltages, respectively [2]. As for the inter-channel spectral overlapping, the DSDD is advantageous over the first two techniques mentioned above, as the approach of filtration by metal foils can be employed in each source-detector assembly independently and thus offers more freedom in spectral channelization. Yet, the shaping of spectral channelization in spectral CT based on the DSDD technology is extremely challenging, due to the requirement that an optimal spectral channelization should be able to freely (continuously) manipulate each channel's spectral center and boundary, which may demand the materials that just do not exist. Moreover, the DSDD technique limits the material decomposition to be carried out in the image domain (post-reconstruction), which, again, may not be able to make full use of the spectral information carried in projection data to reach the performance that may ultimately be achieved [8, 11].

In photon-counting CT, a spectral channel is shaped by energy thresholding and thus the spectral channelization in data acquisition becomes much more feasible [3–5]. Acquisition of raw projection data is carried out in spectral channels of low and high energy simultaneously so that the spatial misregistration and vulnerability to involuntary visceral motion that exist in the switching of peak voltages or dual-source technology can be avoided. Meanwhile, virtually all electronic noise can be eliminated via energy thresholding at the low end of energy and thus substantially improves the image quality at low-dose situations [12, 13]. Moreover, the image formation via one-step reconstruction algorithms is anticipated to deliver further better image quality [14–16]. Altogether, these technological advantages over its energy-integration counterpart are encouraging the imaging community to have a high expectation on the clinical utility of spectral imaging in photon-counting CT [3–5].

In addition to the technological advantages just mentioned, more sophisticated schemes in spectral channelization, such as those with inter-channel overlapping and gapping, are possible via energy thresholding. By taking statistic sufficiency as the figure of merit (FOM), it has been shown that gapping in spectral channelization can lead to improved image quality in material-specific imaging [17]. In a simulation study, the improvement in the contrast-to-noise ratio (CNR) between a target (calcium) and background (water) due to gapping in spectral channelization (and thus reduction of correlation in acquired projection data) was demonstrated in dual-energy projection (radiographic) imaging, under the awareness that a gapping in spectral channelization may not be reasonable in terms of radiation dose efficiency [17]. Perhaps just due to this awareness, to the best of our knowledge, the potential improvement in CNR that can be offered by the gapping in spectral channelization in spectral CT has not been studied in-depth until recently [18, 19]. Hence, our first task in this chapter is to present the data obtained in an in-depth investigation into the effect of inter-channel gapping on the performance of spectral (material-specific and virtual monochromatic) imaging in photon-counting spectral CT, with an emphasis on the trade-off between the improvement in image quality for differentiating soft tissue and the penalty in dose efficiency.

Four spectral channels are presently available in current photon-counting x-ray detectors [20, 21] and can be employed to support 4-MD (material decomposition)-based spectral imaging in photon-counting CT for potential clinical applications, wherein up to two K-edge materials (e.g., iodine and gadolinium) can be accommodated to boost the contrast, by making use of the physiological effect of blood compartment. Even more spectral channels, e.g., eight, are available in some photon-counting x-ray detectors [4], but, as reported by us and others in the literature [22–24], 2-MD implemented with two spectral channels is usually sufficient for spectral imaging of the biological tissues/organs in the human body. Hence, as for the task of imaging biological materials, the additional spectral channels can be used to implement sophisticated schemes, e.g., the interleaved spectral channels to be presented in this chapter. As such, in addition to avoidance of wasting x-ray photons in the gapped spectral channelization (all the x-ray photons that have penetrated the human body should be used for image formation), an improvement in image quality is anticipated (spectral overlapping is eradicated in the interleaved spectral channelization and thus reduces the inter-channel correlation in projection data) in the interleaved cases. Hence, our second task in this chapter is to show the performance of 2-MD-based spectral imaging of biological tissues (soft tissues and bone) in photon-counting CT with data acquisition carried out in interleaved spectral channelization, by exhausting all the possible interleaved schemes.

In this chapter, through a simulation study using an anthropomorphic head phantom that mimics the intracranial soft tissues and bony structures and an image quality phantom, we present the data acquired in the process of accomplishing the two tasks. Cadmium zinc telluride is specifically assumed as the material for x-ray photon-counting detector under both ideal and realistic detector spectral responses [25]. The CNR gauged over targeted regions of interest (ROIs) between soft tissues is adopted as the FOM, while the benchmark scheme of spectral channelization (two adjacent spectral channels) for 2-MD-based spectral imaging is taken as the reference. The material decomposition is carried out in projection domain, while the evaluation of image quality goes over both material-specific and virtual monochromatic imaging (VMI). The materials presented in this chapter can extend our understanding of the physical foundation of 2-MD-based spectral imaging in photon-counting CT and other x-ray-based imaging modalities and provide information on the design and instrumentation of spectral channelization schemes for optimal image quality and dose efficiency in spectral imaging of biological tissues.

2 Signal Detection, Material Decomposition, and Spectral Imaging in Photon-Counting CT

The modeling of signal detection and carrying out of material decomposition in photon-counting CT has been detailed in our recent publications [19, 22, 26–28]. If the noise in data acquisition is taken into consideration, then under certain conditions, the detection of signal in photon-counting CT can be analytically characterized by

$$
I_k(L) = \text{Poisson} \left(\int_{E_{\min}}^{E_{\max}} D_k(E) \, N_0(E) \exp \left(-\int_L \mu (x; E) \, dl \right) dE \right),
\tag{1}
$$

where L denotes the X-ray path and the subscript k indexes the spectral channel used in data acquisition. Poisson($\cdot$) defines the numerical operation that generates random numbers observing the Poisson distribution. $N_0(E)$ is the spectrum of X-ray source and that of a typical X-ray tube used in CT is given in Fig. 1a, in which the photons with energy lower than 20 keV are removed by intrinsic and external filtration that is commonly used in practice. $D_k(E)$ takes into account the detector's efficiency $\eta(E)$ and response $S_k(E)$ in the kth spectral channel [19, 22, 26–28], i.e.,

$$
D_k(E) = S_k(E)\eta(E).
\tag{2}
$$

By definition,

$$
S_k(E) = \int_{E_{min}^k}^{E_{\max}} R\left(E, E'\right) dE',
\tag{3}
$$

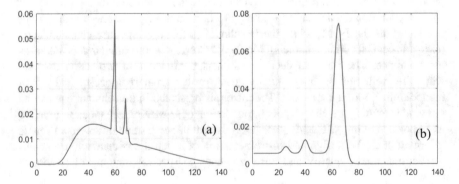

Fig. 1 Spectral profiles characterize (**a**) a typical CT x-ray tube and (**b**) the spectral response of a CZT photon-counting detector (units of x- and y-axis: keV and 1.7×10^{10} photons/cm^2·s)

where E denotes the incident energy and $\left[E_{\min}^k, E_{\max}^k\right]$ the kth spectral channel. The spectral response of an ideal detector $R(E, E')$ is assumed as a pulse (or delta) function, whereas that of a realistic detector is assumed as a summation of Gaussian functions. Figure 1b exemplifies the spectral response of photon-counting detector made of cadmium zinc telluride (CZT), wherein the effects of charge-sharing, Compton scatter, and fluorescent-escaping are modeled [20, 25]:

$$R\left(E, E'\right) = c_1(E) \left(\exp\left(-(E - E')^2/2\sigma_1^2\right) + c_f(E) \exp\left(-(E - E_e)^2/2\sigma_2^2\right)\right.$$
$$\left. + c_2(E) \exp\left(-(E - E' + E_e)^2/2\sigma_2^2\right) + B\left(E, E'\right)\right).$$
(4)

In material decomposition, one can write

$$\mu\left(\boldsymbol{x}; E\right) = \sum_{i=1}^{K} a_i\left(\boldsymbol{x}\right) \mu_i(E).$$
(5)

Subsequently, Eq. (1) can be turned into

$$I_k(L) = \text{Poisson}\left(\int_{E_{\min}}^{E_{\max}} D_k(E)\, N_0(E) \exp\left(-\sum_{p=1}^{P} A_p(L)\mu_p(E)\right) dE\right).$$
(6)

Letting $p = 1, 2, \ldots, P$, $A_p(L) = \int_L a_p(\boldsymbol{x})dl$ denotes the line integral of $a_p(\boldsymbol{x})$ associated with the mass distribution of material p, which can be obtained from $I_k(L)$ ($k = 1, 2, \ldots, K$) by solving the integral equations in Eq. (6) via numerical methods, e.g., the iterative Newton-Raphson algorithm, in which the integral equations' initial condition is obtained by system calibration via polynomial data fitting [19, 22, 26–28]. The basis (material-specific) images $a_p(\boldsymbol{x})$ ($p = 1, 2, \ldots, P$) can be reconstructed from $A_p(L)$ ($p = 1, 2, \ldots, P$) using FBP algorithms [29]. Then, a virtual monochromatic image can be generated by setting the energy E in Eq. (5), while virtual monochromatic analysis can be carried out by varying E over a range that is adequate to the imaging task.

As mentioned above, four spectral channels are commonly available in the photon-counting x-ray detectors for implementation of photon-counting spectral CT, which is assumed for our study in this work. All the potential schemes in spectral channelization are tabulated in Table 1 and briefly delineated below.

3 Spectral Channelization for Data Acquisition in 2-MD-Based Spectral Imaging

3.1 Benchmark Spectral Channelization

Corresponding to the spectrum of a typical x-ray tube used in CT, the basic and most straightforward (called benchmark hereafter) scheme of spectral channelization ([18 58] [59 140] keV) for 2-MD-based spectral imaging is displayed in Fig. 2a.

Table 1 Definition of the spectral channelization schemes for 2-MD-based spectral imaging in photon-counting CT

	Schemes	Spectrum	Gap/interleave	Channel's EE (keV)	Data merge
Benchmark	$((ch_1 + ch_2),$ $(ch_3 + ch_4))$	Full	No	(42.4, 78.3)	Pre-MD
1	(ch_1, ch_2)	Half	No	(34.2, 51.3)	No
2	(ch_2, ch_3)	Half	No	(51.3, 63.9)	No
3	(ch_3, ch_4)	Half	No	(63.9, 92.9)	No
4	(ch_1, ch_3)	Half	Gapped	(34.2, 63.9)	No
5	(ch_2, ch_4)	Half	Gapped	(51.3, 92.9)	No
6	(ch_1, ch_4)	Half	Gapped	(34.2, 92.9)	No
7	$((ch_1 + ch_3),$ $(ch_2 + ch_4))$	Full	Interleaved	(48.7, 72.3)	Pre-MD
8	$((ch_1 + ch_4),$ $(ch_2 + ch_3))$	Full	Interleaved	(62.6, 57.7)	Pre-MD
9	$(ch_1,$ $ch_3) + (ch_2,$ $ch_4)$	Full	Interleaved	(34.2, 63.9), (51.3, 92.9)	Post-recon
10	$(ch_1,$ $ch_4) + (ch_2,$ $ch_3)$	Full	Interleaved	(34.2, 92.9), (51.3, 63.9)	Post-recon

EE effective energy, defined as the spectral centroid of each channel, *Post-Recon* post reconstruction

The four spectral channels ([18 43], [44 58], [59 72], and [73140] keV) under ideal and realistic detector response are displayed in Fig. 2b, c, respectively. The criterion for spectral channelization is to make sure that the photon counts in each channel be roughly equal prior to their passing through the object, though other criteria are reported in the literature [30]. Figure 2c illustrates how the distortion in detector's spectral response (Fig. 1b) may lead to inter-channel overlapping in spectral channelization and altered spectral shape. Readers who wish to know more details about detector spectral distortion and its effect on the performance of spectral imaging in photon-counting CT are referred to the Refs. [4, 20, 21, 25].

3.2 Spectral Channelization Using Half Source Spectrum

3.2.1 Spectral Channelization with Two Abutted Channels

Under ideal detector response, the three schemes of spectral channelization with two abutted channels that are respectively tantamount to half source spectrum that can be employed for 2-MD-based spectral imaging in photon-counting CT are illustrated in Fig. 3a–c. For ease in expression, they are designated as (ch_1, ch_2), (ch_2, ch_3), and (ch_3, ch_4), where $(\cdot, \cdot)$ denotes not only the scheme of spectral channelization but

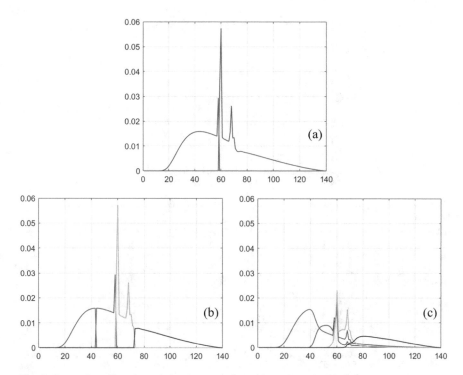

Fig. 2 Spectral profiles characterize the case of two channels under ideal detector response (**a**), and that of four channels under ideal (**b**) and realistic (**c**) spectral responses (units of x- and y-axis: keV and 1.7×10^{10} photons/cm^2·s)

also the process of 2-MD-based spectral imaging, i.e., the formation of material-specific and virtual monochromatic images. Notably, because the two spectral channels are adjacent to each other, the spectral distortion in a realistic detector's response leads to severe inter-channel overlapping (similar to that illustrated in Fig. 2c), though pictorialization is not provided here due to space limitation.

3.2.2 Spectral Channelization with Two Gapped Channels

There exist three schemes, denoted as (ch_1, ch_3), (ch_2, ch_4), and (ch_1, ch_4), in the spectral channelization with two gapped channels that are respectively tantamount to half source spectrum, and the cases that are presented in Fig. 4a–c are modeled under ideal detector response. Again, the cases under realistic response are not provided due to limitation in space. Note that, compared to the cases of abutted channels just mentioned, only very minor, if any, inter-channel overlapping exists, because the inter-channel gap is relatively large.

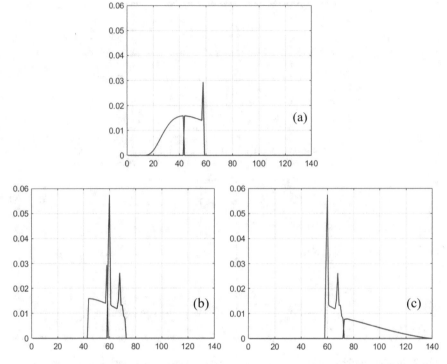

Fig. 3 Spectral profiles characterize the channelization schemes under ideal detector response with two adjacent channels using half source spectrum only: (**a**) (ch$_1$, ch$_2$), (**b**) (ch$_2$, ch$_3$), and (**c**) (ch$_3$, ch$_4$) (units of x- and y-axis: keV and 1.7×10^{10} photons/cm^2·s)

3.3 Spectral Channelization Using Full Source Spectrum

Given four spectral channels, gapping and interleaving can be manipulated in spectral channelization, with data merge being carried out either pre- or post-reconstruction. Below is a brief delineation of each of these schemes.

3.3.1 Interleaved Spectral Channelization and Pre-reconstruction Data Merge

As presented in Fig. 5a, b, two cases exist in this category of channelization under ideal detector response. The first merges the projection data acquired in channels 1 and 3 (ch$_1$ + ch$_3$) and channels 2 and 4 (ch$_2$ + ch$_4$), and the second merges the data in channels 1 and 4 (ch$_1$ + ch$_4$) and channels 2 and 3 (ch$_2$ + ch$_3$), followed by 2-MD and image reconstruction. For clarity in illustration, the cases in Fig. 5 correspond to ideal detector response only. However, it should be easy for us to understand that the inter-channel overlapping in these two cases is effectively severe.

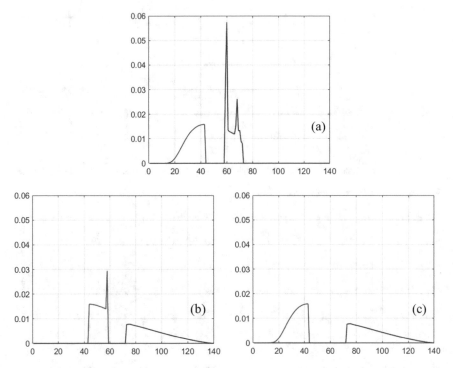

Fig. 4 Spectral profiles characterize the data acquisition in two gapped channels using half source spectrum only: (**a**) (ch_1, ch_3), (**b**) (ch_2, ch_4), and (**c**) (ch_1, ch_4) (units of x- and y-axis: keV and 1.7×10^{10} photons/cm^2·s)

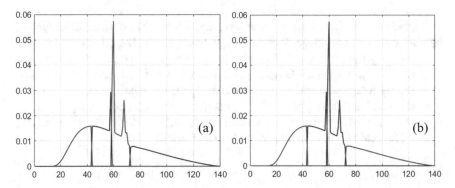

Fig. 5 Spectral profiles characterize the data acquisition via interleaved channels: (**a**) (($ch_1 + ch_3$), ($ch_2 + ch_4$)) and (**b**) (($ch_1 + ch_4$), ($ch_2 + ch_3$)) (units of x- and y-axis: keV and 1.7×10^{10} photons/cm^2·s)

3.3.2 Interleaved Spectral Channelization and Post-reconstruction Data Merge

There are also two cases in this category of spectral channelization. As shown in the top row of Fig. 6, 2-MD and image formation are carried out with the data acquired

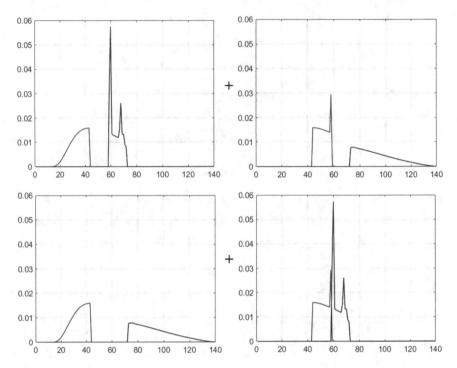

Fig. 6 Spectral profiles characterize the data acquisition via interleaved channels: (top) (ch_1, ch_3) + (ch_2, ch_4), (bottom) (ch_1, ch_4) + (ch_2, ch_3) (units of x- and y-axis: keV and 1.7×10^{10} photons/cm^2·s)

in channels 1 and 3 (ch_1, ch_3) and channels 2 and 4 (ch_2, ch_4), respectively, followed by data merge in the image domain (image summation). The other case is at the bottom of Fig. 6, in which 2-MD and image formation are carried out with the data acquired in channels 1 and 4 (ch_1, ch_4) and channels 2 and 3 (ch_2, ch_3), respectively, followed by image summation. Again, due to space limitation, the cases in Fig. 6 are under ideal detector response only. Notably, the inter-channel overlapping in the scheme (ch_1, ch_3) + (ch_2, ch_4) is actually mild, whereas that in the scheme (ch_1, ch_4) + (ch_2, ch_3) (mainly in (ch_2, ch_3)) is in fact severe.

3.4 Scan Techniques, Basis Materials, and Phantom for Image Quality Assessment

In the simulation study supported by the CT simulator modified from its prototype [31], the photon-counting CT is assumed to work at 140 kVp, 1000 mA, and 1 rot/s gantry rotation speed. The CT is configured with a CZT photon-counting detector that is a curved array at 864×16 dimension and 1.024×1.092 mm^2 pitch,

Fig. 7 A transverse view of
the phantom and the ROIs
(target and background) for
CNR measurement

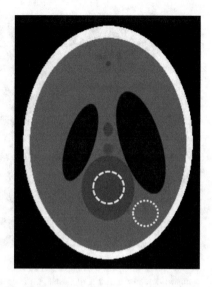

with the source-to-iso and source-to-detector distances are 541.0 and 949.0 mm, respectively. Specifically, the architecture of the photon-counting CT is similar to a typical clinical multi-detector CT for diagnostic imaging.

Without losing generality, soft tissue and cortical bone are chosen as the basis materials for 2-MD-based spectral imaging. The modified Shepp-Logan phantom (Fig. 7) is used to assess the imaging performance associated with the spectral channelization schemes. Two regions of interest (ROIs) are defined as the target (dashed circle) and background (dotted circle) for gauging contrast-to-noise ratio. The mass attenuation coefficients of soft tissues and cortical bone in the phantom and their variation over energy are determined by consulting authoritative publications [32–34] and more details can be found in the Refs. [19, 22, 26–28].

4 Image Quality Evaluation and Verification

4.1 Material Decomposition and Image Formation in Benchmark Spectral Channelization

Initially, we check the feasibility of all the spectral channelization schemes tabulated in Table 1 and found that, except for the two schemes (ch_1, ch_2) and $((ch_1 + ch_4), (ch_2 + ch_3))$, all cases listed in Table 1 can successfully carry out material decomposition and generate material-specific images for virtual monochromatic imaging at acceptable accuracy (i.e., no artifact). The major reason underlying the inability of scheme (ch_1, ch_2) is that the energy of the x-ray photons in channels 1 and 2 is too low to penetrate the head phantom for signal detection, while that

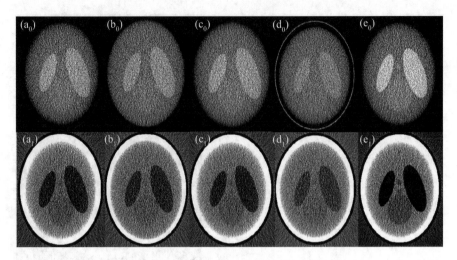

Fig. 8 Material-specific images (soft tissue, top row; cortical bone, bottom row) acquired under realistic detector spectral response, corresponding to the benchmark $((ch_1 + ch_2), (ch_3 + ch_4))$ and other schemes (ch_2, ch_4), (ch_1, ch_4), $((ch_1 + ch_3), (ch_2 + ch_4))$, and $(ch_1, ch_3) + (ch_2, ch_4)$ (left to right)

associated with the scheme $((ch_1 + ch_4), (ch_2 + ch_3))$ is that the effective energy of the two synthesized spectral channels are too close to each other (57.7 keV vs. 62.6 keV; see Table 1), leading to uncertainty (poor conditioning in spectral channelization [27]) in material decomposition.

Corresponding to soft tissue and cortical bone, the material-specific images acquired under realistic detector spectral response are displayed in the top and bottom rows of Fig. 8. Consistent to clinical practice, the display window level (WL) of the images is set at the intensity of brain parenchyma, while the window width (WW) is eight times the noise in the parenchymatic area. From left to right, the images in the first column correspond to the benchmark spectral channelization (item 0 in Table 1), while the others in turn correspond to the schemes (ch_2, ch_4), (ch_1, ch_4), $((ch_1 + ch_3), (ch_2 + ch_4))$, and $(ch_1, ch_3) + (ch_2, ch_4)$ (items 5, 6, 7, and 9 in Table 1). Note that the schemes (ch_2, ch_4) and (ch_1, ch_4) are in the category gapped spectral channels, while the schemes $((ch_1 + ch_3), (ch_2 + ch_4))$ and $(ch_1, ch_3) + (ch_2, ch_4)$ are interleaved spectral channels. Visual inspection of the images in Fig. 8 tells us that both the material decomposition and image formation have been simulated successfully.

Generated from material-specific images, the VMIs acquired under realistic detector spectral response at 18, 35, 50, 65, and 80 keV are shown in Fig. 9, in the order the same as their counterparts in Fig. 8. The display window at each individual energy level is adequately set (WL, the level of brain parenchyma; WW, eight times the noise of parenchyma) to ensure that the visual perceiving of noise over the energy levels is almost identical. Visual inspection of the images in Fig. 9 tells us that virtual monochromatic imaging has been simulated successfully. Additionally, we conduct a spot check by plotting the profiles in Fig. 10 along the dotted lines corresponding to the case $(ch_1, ch_3) + (ch_2, ch_4)$ at 18, 50, and 80 keV (bottom row in Fig. 9), demonstrating the accuracy of virtual monochromatic imaging under complicated spectral channelization scheme.

4.2 Material-Specific Imaging in Interleaved/Gapped Spectral Channelization

Under both ideal and realistic detector responses, we quantitatively evaluate the performance of material-specific imaging. Along with the quantitative results listed in Table 2, the contrast-to-noise ratio over the nine feasible spectral channelization schemes is plotted in Fig. 11. The chart in the left panel corresponds to the material-specific images of soft tissue, while that on the right is obtained from those images of cortical bone. It is noted in the material-specific image of soft tissue that the schemes (ch_1, ch_4) and $(ch_1, ch_3) + (ch_2, ch_4)$ outperform the benchmark scheme $((ch_1 + ch_2), (ch_3 + ch_4))$, while the schemes (ch_2, ch_4), (ch_1, ch_3), and $(ch_1, ch_4) + (ch_2, ch_3)$ perform moderately worse than the benchmark. Similar behavior in the contrast-to-noise ratio is observed in the material-specific images of cortical bone. A further note can be made on the contrast-to-noise ratio corresponding to the realistic detector response that it is considerably lower than those of the ideal ones, in almost all the schemes of spectral channelization.

4.3 Virtual Monochromatic Imaging in Interleaved/Gapped Spectral Channelization

Our quantitative evaluation of the performance in VMI also runs through the cases under both ideal and realistic detector responses. The contrast-to-noise ratio in VMI corresponding to the nine feasible spectral channelization schemes is presented in Fig. 12, wherein the profiles in the left panel correspond to ideal detector spectral response, while those on the right are obtained under realistic detector response. Under the ideal detector spectral response, it is observed that, in terms of the best scenario contrast-to-noise ratio (i.e., the contrast-to-noise ratio at the sweet spot of energy), the scheme $(ch_1, ch_4) + (ch_2, ch_3)$ outperforms the benchmark scheme $((ch_1 + ch_2), (ch_3 + ch_4))$, and the scheme $(ch_1, ch_3) + (ch_2, ch_4)$ performs comparably to the benchmark despite of a moderate shift (roughly 5 keV) to the low-energy end, followed by the scheme $((ch_1 + ch_3), (ch_2 + ch_4))$. Similarly, an inspection of Fig. 12a, b shows that the distortion in realistic detector's spectral response markedly drags down the performance in VMI. Interestingly, note that the best scenario performance of virtual monochromatic imaging associated with the benchmark scheme remains almost unchanged from ideal to realistic detector

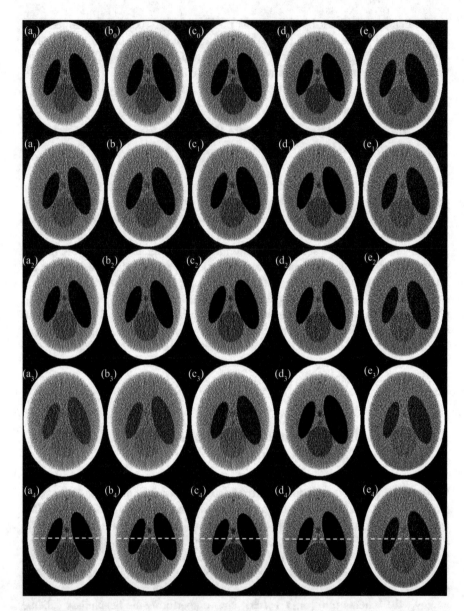

Fig. 9 VMIs acquired under realistic detector spectral response at 18, 35, 50, 65, and 80 keV (left to right) corresponding to the benchmark scheme (($ch_1 + ch_2$), ($ch_3 + ch_4$)) and other schemes (ch_2, ch_4), (ch_1, ch_4), (($ch_1 + ch_3$), ($ch_2 + ch_4$)), and (ch_1, ch_3) + (ch_2, ch_4) (top to bottom)

spectral response. As a result, the benchmark scheme performs the best at its sweet spot, and the scheme (ch_1, ch_4) + (ch_2, ch_3) as well, though both of them diminish slightly in the spectral breadth.

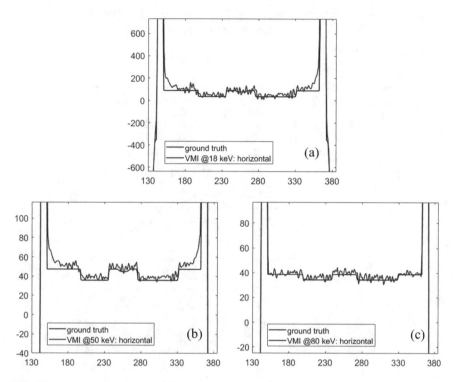

Fig. 10 Intensity profiles plotted along the dashed lines in the bottom of Fig. 9 against their ground truths at (**a**) 18, (**b**) 50, and (**c**) 80 keV, under channelization scheme $(ch_1, ch_3) + (ch_2, ch_4)$ (units of x- and y-axis: pixels and Hounsfield unit)

Table 2 The contrast-to-noise ratio measured in differing spectral channelization schemes under ideal and realistic detector response, respectively

Schemes	CNR (ideal)		CNR (realistic)	
	Soft tissue	Cortical bone	Soft tissue	Cortical bone
(ch_1, ch_3)	0.5312	0.8040	0.4349	0.3864
(ch_1, ch_4)	0.6476	1.0636	0.4544	0.7239
(ch_2, ch_3)	0.2012	0.3929	0.1504	0.2523
(ch_2, ch_4)	0.4629	0.7809	0.3626	0.5667
(ch_3, ch_4)	0.2717	0.4275	0.2271	0.3348
$((ch_1 + ch_2), (ch_3 + ch_4))$	0.5312	0.9362	0.4349	0.7096
$((ch_1 + ch_3), (ch_2 + ch_4))$	0.2347	0.4034	0.2666	0.4346
$(ch_1, ch_3) + (ch_2, ch_4)$	0.5993	1.0973	0.3678	0.5818
$(ch_1, ch_4) + (ch_2, ch_3)$	0.3865	0.7304	0.2801	0.4720

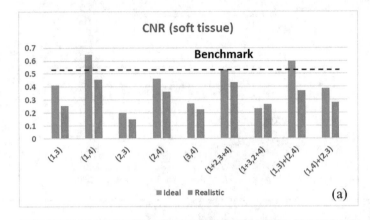

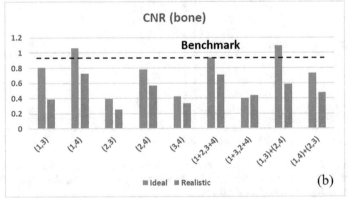

Fig. 11 Contrast-to-noise ratio gauged in material-specific images of soft tissue (**a**) and cortical bone (**b**) corresponding to the nine feasible spectral channelization schemes listed in Table 1

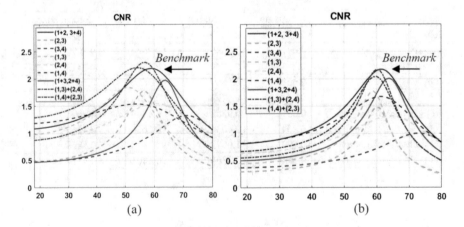

Fig. 12 Variation of contrast-to-noise ratio in VMI over all the feasible spectral channelization schemes under ideal (**a**) and realistic (**b**) detector spectral response (unit of abscissa: keV)

5 Closing Remarks

To fully understand the technological challenges associated with photon-counting CT and fulfill its potential in clinical utility, we need an in-depth revisiting of the underlying physics and investigating the solutions to overcome the challenges. Recently, we have investigated the dimensionality of material space, the conditioning of basis materials (functions) [22, 26] and spectral channelization (energy binning) [26, 27], and the correlation of noise in the material-specific (basis) images [28] as well. In practice, the availability of more than two spectral channels in photon-counting CT offers the opportunity for 2-MD-based spectral imaging via sophisticated spectral channelization schemes. By exhausting all the possible channelization schemes that are implementable with four spectral channels, we presented the results obtained in our studies regarding the feasibility and performance of spectral imaging in photon-counting CT associated with those sophisticated channelization schemes in this chapter. Below are the major points that we want to make prior to concluding the chapter.

Originally, the spectral imaging in DECT was proposed under the assumption of acquiring projection data at two distinct energy levels [8]. Alvarez and Macovski showed in their milestone paper that, via data fitting (or system calibration) techniques, the spectral imaging in DECT can be implemented using a polychromatic x-ray source at two different peak voltages. Hence, as expected, the gapped spectral channelization (ch_1, ch_3), (ch_2, ch_4), and (ch_1, ch_4) can be utilized to carry out spectral imaging in photon-counting CT, despite the existence of inter-channel gapping in the spectrum. It is also within expectation that the interleaved spectral channelization with post-reconstruction data merge $((ch_1, ch_4) + (ch_2, ch_3)$ and $(ch_1, ch_3) + (ch_2, ch_4)$; see Sect. 3.3.2) works well, as basically there is no change in the process of material decomposition and image formation. Notably, however, the interleaved spectral channelization with pre-reconstruction data merge $(((ch_1 + ch_3), (ch_2 + ch_4))$; see Sect. 3.3.1) also works for spectral imaging in photon-counting CT, though in principle it should still be within anticipation. This is a novel finding and may be of practical relevance from our point of view.

Interestingly, the channelization scheme (ch_1, ch_4) outperforms the benchmark $((ch_1 + ch_2), (ch_3 + ch_4))$ in material-specific imaging, though only half radiation dose is used in the former for data acquisition and image formation. The rationale may be most likely that the role played by inter-channel correlation is dominant in determining the performance of material-specific (or basis material) imaging. Notably, the inter-channel correlation here should be perceived twofold: (i) the inter-channel spectral overlapping, i.e., the spectral conditioning [27] or simply the difference between the effective energies of two spectral channels in data acquisition [9, 10] (see Table 1), and (ii) the similarity in basis materials' attenuating property over spectral channels (i.e., the conditioning of basis materials [21]).

The ranking of performance in virtual monochromatic imaging differs from that in material-specific imaging. As noted, the scheme (ch_1, ch_4), which is the best in material-specific imaging, performs markedly worse than the benchmark

$((ch_1 + ch_2), (ch_3 + ch_4))$ in VMI. However, the scheme $(ch_1, ch_4) + (ch_2, ch_3)$, which performs markedly worse than the benchmark in material-specific imaging, is the best in virtual monochromatic imaging. Hence, it would be reasonable to choose different spectral channelization schemes that are optimal for material-specific imaging and VMI, respectively, if more than two spectral channels are available for spectral imaging in photon-counting CT. Also noted is the fact that, in virtual monochromatic imaging (Fig. 12), all the schemes using full x-ray source spectrum (and thus all x-ray photons) outperforms those using half spectrum (and thus half x-ray photons) only. Thus, fairly speaking, though the correlation in noise corresponding to each material-specific image plays a role, the total number of x-ray photons used for data acquisition and image formation plays a pivotal role in determining the performance of virtual monochromatic imaging.

In virtual monochromatic imaging, it is noted that the location at which the sweet spot is (i.e., highest CNR is reached) varies over the schemes of spectral channelization. Visually, Fig. 12 shows that variation of the sweet spot locations correlates with the effective energy of each channelization scheme. It is also observed that the spectral breadth of each bell-shaped CNR profile correlates with the effective breadth of the spectral channelization used in data acquisition. For example, all the schemes using full spectrum $(((ch_1 + ch_2), (ch_3 + ch_4))$, $((ch_1 + ch_3), (ch_2 + ch_4)), (ch_1, ch_3) + (ch_2, ch_4)$, and $(ch_1, ch_4) + (ch_2, ch_3))$ are relatively broad in spectral distribution compared to those using only half spectrum $((ch_1, ch_3), (ch_2, ch_3), (ch_2, ch_4)$, and $(ch_3, ch_4))$, and this is also the case in the scheme (ch_1, ch_4). Moreover, a comparison between the counterparts in the profiles plotted in Fig. 12a, b tells us that the distortion in a detector's spectral response diminishes the spectral breadth in virtual monochromatic imaging.

Only the two-step approach (i.e., material decomposition + image reconstruction) is used in the simulation study, but the gapped/interleaved spectral channelization schemes are readily applicable in the case wherein the one-step approach [14–16] is employed for material decomposition and image formation, in which better performance in spectral imaging is expected. Moreover, only four spectral channels are considered in the study to exemplify the feasibility of spectral imaging in photon-counting CT via gapped/interleaved spectral channels. It is surely possible to have more spectral channels, e.g., currently the largest number reported in the literature is eight, for spectral channelization. However, it is a system designer's discretion in making the decision if more spectral channels should be shaped and used for spectral imaging of biological tissues in photon-counting CT, in light of the possibility of photon starvation in each individual spectral channel that may damage or ruin the feasibility of material decomposition and thus spectral imaging proportionally increases with the number of spectral channels.

An inspection of Figs. 11 and 12 tells us that the spectral distortion in a realistic detector's spectral response degrades the performance of spectral imaging in photon-counting CT considerably. Specifically, in virtual monochromatic imaging, due to spectral distortion, the interleaved scheme $(ch_1, ch_4) + (ch_2, ch_3)$ loses its advantage against the benchmark $((ch_1 + ch_2), (ch_3 + ch_4))$, while the other interleaved scheme $(ch_1, ch_3) + (ch_2, ch_4)$, which performs as well as the benchmark under ideal detector spectral response, is left behind by the benchmark. Therefore, reduction of distortion in detector's spectral response is always desirable, in order to get optimal performance of spectral imaging in photon-counting CT.

Only the head is considered as the anatomic site in the simulation study, but the study can be readily extended to other anatomic locations, such as the thorax, abdomen, and pelvis, in which the spectral channelization needs to be optimized to be adequate for the anatomic locations at larger dimensions. Also, only the contrast-to-noise ratio is adopted as the figure of merit for assessment of imaging performance in this chapter, which is a limitation in light of the fact that other figure of merits, such as those under observer or model observer studies, may be of higher fidelity in image quality assessment [35–40]. However, we'd like to indicate that the contrast-to-noise ratio should be sufficient for the purpose of the simulation study presented in this chapter, since all the cases in our study are under identical x-ray source spectrum and single CZT detector material [24], i.e., an intra-class comparison study. Up to date, only very limited numbers of materials, CZT, cadmium telluride, and silicon, have been developed for photon-counting x-ray detector. We think that the figure of merits supported by observer or model observer studies [34–39] may have to be employed for image quality assessment and comparison across the detector materials in photon-counting CT for spectral imaging of biological tissues [26–28].

Finally, we want to indicate that, in addition to facilitating the implementation of interleaved/gapped spectral channelization for data acquisition, the availability of more than two spectral channels in photon-counting CT may provide the opportunity of implementing spectral (energy) weighting [41, 42] on the projection data in a way that is finer than that in the case wherein only two spectral channels are available. Optimistically, we anticipate the imaging community will invest more effort on the research and development along this technical avenue, with the hope that the integration of spectral (energy) weighting and multispectral channels may strengthen the photon-counting CT's capability of differentiating soft tissues in diagnosis of cardiovascular, oncologic, and neurovascular diseases.

References

1. Chandra N, Langan DA. Gemstone detector: dual energy imaging via fast kVp switching. In: Johnson T, Fink C, Schönberg SO, Reiser MF, editors. Dual energy CT in clinical practice. Berlin, Heidelberg: Springer Berlin Heidelberg; 2011. p. 35–41. https://doi.org/10.1007/174_2010_35.
2. Krauss B, Schmidt B, Flohr TG. Dual source CT. In: Johnson T, Fink C, Schönberg SO, Reiser MF, editors. Dual energy CT in clinical practice. Berlin, Heidelberg: Springer Berlin Heidelberg; 2011. p. 11–20. https://doi.org/10.1007/174_2010_44.
3. Willemink MJ, Persson M, Pourmorteza A, Pelc NJ, Fleischmann D. Photon-counting CT: technical principles and clinical prospects. Radiology. 2018;289(2):293–312.
4. Danielsson M, Persson M, Sjölin M. Photon-counting x-ray detectors. Phys Med Biol. 2021;66:03TR01.

5. Flohr T, Petersilka M, Henning A, Ulzheimer S, Ferda J, Schmidt B, Photon-counting CT. review. Phys Med. 2020;79(November):126–36.
6. U.S. Food & Drug Administration. https://www.fda.gov/news-events/press-announcements/ fda-clears-first-major-imaging-device-advancement-computed-tomography-nearly-decade. Accessed on 5/8/2022.
7. Vlassenbroek A. Dual layer CT. In: Johnson T, Fink C, Schönberg SO, Reiser MF, editors. Dual energy CT in clinical practice. Berlin, Heidelberg: Springer Berlin Heidelberg; 2011. p. 21–34. https://doi.org/10.1007/174_2010_56.
8. Alvarez RE, Macovski A. Energy-selective reconstructions in X-ray computerised tomography. Phys Med Biol. 1976;21(5):733–44.
9. Kelcz F, Joseph PM, Hilal SK. Noise considerations in dual energy CT scanning. Med Phys. 1979;6(5):418–25.
10. Primak AN, Ramirez Giraldo JC, Liu X, Yu L, McCollough CH. Improved dual-energy material discrimination for dual-source CT by means of additional spectral filtration. Med Phys. 2009;36(4):1359–69.
11. Alvarez R, Seppi E. A comparison of noise and dose in conventional and energy selective computed tomography. IEEE Trans Nucl Sci. 1979;26(2):2853–6.
12. Taguchi K, Iwanczyk JS. Vision 20/20: single photon counting x-ray detectors in medical imaging. Med Phys. 2013;40(10):100901.
13. Wehrse E, Klein L, Rotkopf LT, Wagner WL, Uhrig M, Heussel CP, Ziener CH, Delorme S, Heinze S, Kachelriess M, Schlemmer HP, Sawall S. Photon-counting detectors in computed tomography: from quantum physics to clinical practice. Radiologe. 2021;61(Suppl 1):S1–S10.
14. Barber RE, Sidky EY, Schmidt TG, Pan X. An algorithm for constrained one-step inversion of spectral CT data. Phys Med Biol. 2016;61(10):3784–818.
15. Chen B, Zhang Z, Sidky EY, Xia D, Pan X. Image reconstruction and scan configurations enabled by optimization-based algorithms in multispectral CT. Phys Med Biol. 2017;62(22):8763–93.
16. Mory C, Sixou B, Si-Mohanmed S, Boussel L, Rit S. Comparison of five one-step reconstruction algorithms for spectral CT. Phys Med Biol. 2018;63(15):235001 (19 pages).
17. Wang A, Pelc NJ. Sufficient statistics as a generalization of binning in spectral X-ray imaging. IEEE Trans Med Imaging. 2011;30(1):84–93.
18. Yao Y, Wang A, Pelc N. Efficacy of fixed filtration for rapid kVp-switching dual energy x-ray systems. Med Phys. 2014;41(3):031914.
19. Tang X, Ren Y, Xie H. Photon-counting CT via interleaved/gapped spectral channels: feasibility and imaging performance. Med Phys. 2022;49(3):1445–57.
20. Roessl E, Proksa R. K-edge imaging in x-ray computed tomography using multi-bin photon counting detectors. Phys Med Biol. 2007;52(15):4679–96.
21. Schlomka JP, Roessl E, Dorscheid R, et al. Experimental feasibility of multi-energy photon-counting K-edge imaging in pre-clinical computed tomography. Phys Med Biol. 2008;53(15):4031–47.
22. Tang X, Ren Y. On the conditioning of basis materials and its impact on multi-material decomposition based spectral imaging in photon-counting CT. Med Phys. 2021;48(3):1100–16.
23. Han D, Porras-Chaverri MA, O'Sullivan JA, Politte DG, Williamson JF. Technical note: on the accuracy of parametric two-parameter photon cross-section models in dual energy-CT applications. Med Phys. 2017;44(6):2338–46.
24. Zhang S, Han D, Politte DG, Williamson JF, O'Sullivan JA. Impact of joint statistical dual-energy CT reconstruction of proton stopping power images: comparison to image- and sinogram-domain material decompositin approaches. Med Phys. 2018;45(5):2129–42.
25. Ehn S. Photon-counting hybrid-pixel detectors for spectral X-ray imaging applications (Doctoral dissertation). 2017. Retrieved from https://mediatum.ub.tum.de/doc/1363593/ 1363593.pdf.

26. Ren Y, Xie H, Long W, Yang X, Tang X. Optimization of basis material selection and energy binning in three material decomposition for spectral imaging without contrast agents in photon-counting CT. SPIE Proc. 2020;11312:113124X (8 pages). https://doi.org/10.1117/12.2549678.
27. Ren Y, Xie H, Long W, Yang X, Tang X. On the conditioning of spectral channelization (energy binning) and its impact on multi-material decomposition based spectral imaging in photon-counting CT. IEEE Trans Biomed Eng. 2021;68(9):2678–88.
28. Tang X, Ren Y. Noise correlation in multi-material decomposition-based spectral imaging in photon-counting CT. In: Proceedings of the 16th international meeting on fully three-dimensional image reconstruction in radiology and nuclear medicine, Leuven Belgium, July 19–23; 2021. p. 414–9.
29. Buzug TM. Computed tomography: from photon statistics to modern cone-beam CT. 1st ed. Berlin Heidelberg: Springer; 2008. https://doi.org/10.1007/978-3-540-39408-2.
30. Chen H, Xu C, Persson M, Danielsson M. Optimization of beam quality for photon-counting spectral computed tomography in head imaging: simulation study. SPIE J Med Imaging. 2015;2(4):043504 (16 pages).
31. De Man B, Basu S, Chandra N, Dunham B, Edic P, Iatrou M, McOlash S, Sainath P, Shaughnessy C, Tower B, Williams E. CatSim: a new computer assisted tomography simulation environment. SPIE Proc. 2007;6510:65102G (8 pages). https://doi.org/10.1117/12.710713.
32. Woodard HQ, White DR. The composition of body tissues. Br J Radiol. 1986;59(12):1209–19.
33. White DR, Griffith RV, Wilson IJ. ICRU report 46: Photon, electron, proton and neutron interaction data for body tissues. J Int Comm Radiat Units Meas. 1992;os24(1)
34. Cullen DE, Hubbell JH, Kissel L. EPDL97: the evaluated photon data library. Lawrence Livermore National Laboratory Report UCRL-50400, vol 6, rev 5. 1997.
35. Wagner RF, Brown DG. Unified SNR analysis of medical imaging systems. Phys Med Biol. 1985;30(6):489–518.
36. Metz CE, Wagner RF, Doi K, Brown DG, Nishikawa RM, Myers K. Toward consensus on quantitative assessment of medical imaging systems. Med Phys. 1995;22(7):1057–61.
37. Popescu LM, Myers KJ. CT image assessment by low contrast signal detectability evaluation with unknown signal location. Med Phys. 2013;40(11):111908 (10 pages).
38. Yu L, Leng S, Chen L, Kofler JM, Caryer RE, McCollough CH. Prediction of human observer performance in a 2-alternative forced choice low-contrast detection task using channelized Hotelling observer: impact of radiation dose and reconstruction algorithms. Med Phys. 2013;40(4):041908 (9 pages).
39. Vaishnav JY, Jung WC, Popescu LM, Zeng R, Myers K. Objective assessment of image quality and dose reduction in CT iterative reconstruction. Med Phys. 2014;41(7):071904 (12 pages).
40. Yu L, Chen B, Kofler JM, Favazza CP, Leng S, Kupinski MA, McCollough CH. Correlation between a 2D channelized Hotelling observer and human observers in a low-contrast detection task with multislice reading in CT. Med Phys. 2017;44(8):3990–9.
41. Tapiovaara MJ, Wagner RF. SNR and DQE analysis of broad spectrum x-ray imaging. Phys Med Biol. 1985;30(6):519–29.
42. Cahn RN, Cederstrom B, Danielsson M, Hall A, Lundqvist M, Nygren D. Detective quantum efficiency dependence on x-ray energy weighting in mammography. Med Phys. 1999;26(12):2680–3.

One-Step Basis Image Reconstruction in Spectral CT Based on MAP-EM Algorithm and Polar Coordinate Transformation

Zhengdong Zhou

1 Introduction

The concept of spectral CT was firstly proposed by Alvarez and Macovski in 1976 [1]. Spectral CT has been attracting great research interest in the last decades. Nowadays, spectral dual-energy CT system has been available in various clinical applications, and spectral photon-counting CT system is the research focus in the field of CT imaging. Using spectral CT, virtual monoenergy images can be reconstructed to reduce the effect of beam-hardening artifacts [2–4]. By identifying the energy information with different energy spectra, spectral CT can also provide more material properties and is thus superior to traditional CT in material identity [5, 6]. Spectral CT has broad application prospects in the fields of medical diagnosis, non-destructive testing, security inspection, etc.

Spectral CT has the distinctive ability of material decomposition and identification. Currently, the most widespread methods of material decomposition can be divided into two categories: two-step and one-step methods. Two-step methods include image-domain [7–10] and projection-domain material decomposition methods [1, 11–14]. Image-based methods perform reconstruction first and then decomposition. The image-based methods often integrate some form of empirical beam-hardening correction but require knowledge of the material volumes in advance for perfect beam-hardening correction. On the other hand, projection-based methods perform decomposition first and then reconstruction. However, it is hard to achieve a robust decomposition for typical choices of basis materials (e.g., water, bone, and a high-Z contrast agent) without enough difference of the normalized attenuation profiles; thus, aberrant material line integrals may be caused due to the

Z. Zhou (✉)
State Key Laboratory of Mechanics and Control of Mechanical Structures, Nanjing University of Aeronautics and Astronautics, Nanjing, China
e-mail: zzd_msc@nuaa.edu.cn

© The Author(s), under exclusive license to Springer Nature Switzerland AG 2023
S. Hsieh, K. (Kris) Iniewski (eds.), *Photon Counting Computed Tomography*,
https://doi.org/10.1007/978-3-031-26062-9_10

inevitable statistical noise on photon counts, resulting in strong streak artifacts in the reconstructed images. In two-step methods, loss of information is inevitable, because the first step cannot provide a one-to-one mapping between inputs and outputs. Therefore, the inaccuracy of material decomposition is unavoidable since the second step cannot compensate for the loss of information that occurred in the first step [15]. The one-step method, also named as "one-step inversion," refers to the direct iterative material decomposition method, which performs decomposition and reconstruction simultaneously [16]. The one-step method can overcome the inherent shortcomings of the two-step method and improve the accuracy of material decomposition. In 2013, Cai et al. [17] proposed a one-step material decomposition method based on Bayesian model with Huber prior knowledge, in which the conjugate gradient algorithm is adopted for optimization. Thereafter, several direct iterative projection models have been proposed one after another, and all the methods for solving the models involved complex non-convex function optimization problem [16, 18–21]. Foygel et al. [16] proposed a primal-dual algorithm for material decomposition with one-step inversion from spectral CT projection data, where the image constraints are enforced on the basis maps during the inversion. Combined with a convex-concave optimization algorithm and a local upper bounding quadratic approximation, the algorithm is derived to generate descent steps for non-convex spectral CT data discrepancy terms. Long et al. [18] proposed a penalized-likelihood (PL) method to reconstruct multi-material images using a similar constraint from sinogram data, where the edge-preserving regularizers for each material are employed. And an optimization transfer method with a series of pixel-wise separable quadratic surrogate (PWSQS) functions was developed to monotonically decrease the complicated PL cost function. Weidinger et al. [19] proposed a dedicated statistical algorithm based on local approximations of the negative logarithmic Poisson probability function to perform a direct material decomposition for photon counting spectral CT. The algorithm allows for parallel updates of all image pixels, which can compensate for the rather slow convergence that is intrinsic to statistical algorithms. Mechlem et al. [20] proposed an algorithm based on a semi-empirical forward model for joint statistical iterative material image reconstruction, where the semi-empirical forward model is tuned by calibration measurements. This strategy allows to model spatially varying properties of the imaging system without requiring detailed prior knowledge of the system parameters. And an efficient optimization algorithm based on separable surrogate functions is employed to accelerate convergence and reduce the reconstruction time. Fang et al. [21] proposed an iterative one-step inversion material decomposition algorithm with a Noise2Noise prior. The algorithm estimated material images directly from projection data and used a Noise2Noise prior for denoising. All the methods need to be solved iteratively, and currently no analytical inversion formula is available for the material decomposition problem, let alone for one-step inversion [15].

In previous studies, we proposed methods for spectral CT image reconstruction by using maximum a posteriori expectation maximization (MAP-EM) algorithm [22, 23], which has advantages in noise robustness and image reconstruction quality. Based on previous studies, a novel and robust one-step basis material image reconstruction based on maximum a posteriori expectation-maximization algorithm (MAP-EM-DD) is proposed. Furthermore, by incorporating polar coordinate transformation into MAP-EM-DD, MAP-EM-PT-DD method is proposed. The iterative formulas of MAP-EM-DD and MAP-EM-PT-DD methods are derived. For readability, we present the case of dual-energy CT (DECT) and decompose on a basis of bone tissue and soft tissue, but all the derivations can be adapted to spectral photon-counting CT. The performance of the proposed methods was evaluated and compared with the image domain material decomposition method based on FBP algorithm (FBP-IDD) [24].

2 Materials and Methods

2.1 MAP-EM Statistical Reconstruction Algorithm

In spectral CT system, noise is inevitable in the projection due to the effects of incoherent scattering, pulse accumulation, and electronic noise. The statistical property of the detected data can be described by Poisson model. The MAP-EM algorithm with the Poisson model tries to find an image I to maximize the conditional probability $P(I \mid y)$ by a set of measured projection vector y, as follows [25]:

$$P(I|y) = P(y|I) P(I) / P(y) \tag{1}$$

where $P(I)$ is the prior knowledge of the image and y is the known measured projection. Applying Gibbs prior distribution as the prior knowledge, $P(I)$ can be expressed as follows:

$$P(I) = \exp(-\beta \cdot U(I)) / c \tag{2}$$

where β is a regularization parameter, the constant c is an unknown normalization factor, and the function $U(I)$ denotes the total energy of I.

Applying the form of the Gibbs distribution given by Eq. (2) in the MAP-EM procedure, the iterative formula is given as follows [26]:

$$I_n^{s+1} = \frac{I_n^s}{\sum_k a_{kn} + \beta \frac{\partial U(I^s)}{\partial I_n}} \sum_k \frac{y_k a_{kn}}{\sum_{n'} a_{kn'} I_{n'}^s} \tag{3}$$

where s is the number of iteration, I_n means the nth pixel value in the reconstructed image I, y_k means the projection detected by the kth detector element, a_{kn} represents the probability that the photon passes through the nth pixel collected by the kth detector element, and $\beta \partial U(I^S)/\partial I_n$ is the term of penalty function which can suppress the noise in the reconstructed image. When β equals zero, the algorithm tends toward maximum likelihood expectation-maximization (ML-EM) algorithm.

2.2 Direct Iterative Material Decomposition Method Based on MAP-EM Algorithm

The linear attenuation coefficient of material depends on its specific properties and the incident energy of X-ray. In accordance with the basis material decomposition model, the linear attenuation coefficient of an object can be expressed as the weighted sum of the linear attenuation coefficients of certain materials [27] as follows:

$$\mu(E, l) \approx \sum_{m=1}^{n_f} x_m(l) \mu_m(E) \tag{4}$$

where n_f is the number of basis material, $x_m(l)$ is the basis image of the mth material, l is the path of X-ray, and $\mu_m(E)$ is the linear attenuation coefficient of the mth material at the energy E.

The discrete form of the projection function along path l at certain energy E can be expressed as follows:

$$P(E, l) = \sum_{m=1}^{n_f} \mu_m(E) \boldsymbol{B}_m(l) \tag{5}$$

where $\boldsymbol{B}_m(l) = \int_l x_m(l) dl$ is the projection of the basis image of the mth material along path l.

Let the continuous X-ray energy spectrum under the tube voltage u be divided into Q energy bins. The projection of the target under the tube voltage u can then be expressed as a weighted sum of the projections at each energy bin as follows:

$$P(u, l) = \sum_{q=1}^{Q} w(E_q) \cdot P(E_q, l)$$

$$= \sum_{q=1}^{Q} \sum_{m=1}^{n_f} w(E_q) \mu_m(E_q) \boldsymbol{B}_m(l) = \sum_{m=1}^{n_f} \mu_m(u) \boldsymbol{B}_m(l) \tag{6}$$

where $w(E_q) = n_p(E_q) / \sum_{q=1}^{Q} n_p(E_q)$ is the normalization coefficient of the number of photons in the qth energy bin, $n_p(E_q)$ is the number of photon in the qth energy bin, and $\mu_m(u)$ is the effective linear attenuation coefficient of a material at a given tube voltage u.

The projection of the basis image of the mth material along path l can be expressed as follows:

$$B_m(l) = Ax_m(l), \quad m = 1, \ldots, n_f. \tag{7}$$

Substituting Eq. (7) into Eq. (6), the discrete form of the direct iterative projection function can be obtained as follows [17]:

$$P = JRx \tag{8}$$

Where $x = [x_1, x_2, \ldots, x_{n_f}]^T$ is the set of basis images, the dimension is $\mathcal{R}^{n_f N}$, N is the total number of pixels in the basis images, and J is the joint basis material matrix, which is expressed as follows:

$$J = J_1 \otimes C_K, \quad J_1 = \begin{bmatrix} \mu_1(u_1) & \cdots & \mu_{n_f}(u_1) \\ \vdots & \ddots & \vdots \\ \mu_1(u_{N_e}) & \cdots & \mu_{n_f}(u_{N_e}) \end{bmatrix} \tag{9}$$

where N_e is the total number of energy bins, K is the total observation dimension, C_k is the identity matrix of size $K \times K$, and R is a joint projection operator expressed as follows:

$$R = C_{n_f} \otimes A \tag{10}$$

where C_{n_f} is the identity matrix with size of $n_f \times n_f$, and $\otimes$ denotes the matrix Kronecker product.

For the material decomposition of DECT with two basis materials, i.e., $N_e = 2$, $n_f = 2$, the material decomposition model can be expressed as follows:

$$P = JRx = \begin{bmatrix} \mu_1(u_L) A & \mu_2(u_L) A \\ \mu_1(u_H) A & \mu_2(u_H) A \end{bmatrix} \cdot \begin{bmatrix} x_1 \\ x_2 \end{bmatrix} \tag{11}$$

Where $P = [P_1, P_2]^T$ is the set of projections at two different tube voltages, x_1 and x_2 are the basis images, and $x = [x_1, x_2]^T$ is the set of basis images.

Let $W = JR$, which denotes the transformed projection operator, then $P = Wx$. Evidently, the material decomposition problem is the same as the problem of image reconstruction. In accordance with the transformed projection operator W and the set of projection P, the basis images can be reconstructed by different kinds of reconstruction algorithms.

Let $W = \{c_{kn}\}$, the iterative formulas for the direct material decomposition using MAP-EM algorithm (MAP-EM-DD) can be derived from Eq. (3) as follows:

$$x_{1n}^{s+1} = \frac{x_{1n}^s}{\sum_k c_{kn} + \beta \frac{\partial U(x_1^s)}{\partial x_{1n}}} \sum_k \frac{y_{1k} c_{kn}}{\sum_{n'} c_{kn'} x_{1n'}^s},$$

$$x_{2n}^{s+1} = \frac{x_{2n}^s}{\sum_k c_{kn} + \beta \frac{\partial U(x_2^s)}{\partial x_{2n}}} \sum_k \frac{y_{2k} c_{kn}}{\sum_{n'} c_{kn'} x_{2n'}^s} \qquad (12)$$

where x_{1n} and x_{2n} represent the nth pixel value in basis images x_1 and x_2, respectively; $U\left(x_1^s\right)$ and $U\left(x_2^s\right)$ are the energy functions of basis images x_1^s and x_2^s at sth iteration, respectively; and y_{1k} and y_{2k} are the projection detected by the kth detector element at the two different tube voltages, respectively; c_{kn} represents the weighted probability that the photon passes through the nth pixel collected by the kth detector element.

Furthermore, by applying polar coordinate transformation to replace (x_1, x_2) with (r, θ), we have $x_1 = r \cos \theta$ and $x_2 = r \sin \theta$. Equation (12) can be expressed by Eq. (13) for MAP-EM-PT-DD method as follows:

$$r_n^{s+1} = \frac{r_n^s}{\sum_k c_{kn} + \beta \frac{\partial U(r^s)}{\partial r_n}} \sum_k \frac{y_{1k} c_{kn}}{\sum_{n'} c_{kn'} x_{n'}^s},$$

$$\theta_n^{s+1} = \frac{\theta_n^s}{\sum_k c_{kn} + \beta \frac{\partial U(\theta^s)}{\partial \theta_n}} \sum_k \frac{y_{1k} c_{kn}}{\sum_{n'} c_{kn'} x_{n'}^s} \qquad (13)$$

where $x_{n'}^s = \begin{bmatrix} r \cos \theta \\ r \sin \theta \end{bmatrix}_{n'}^s$, r_n and θ_n represent the value of the nth pixel on the basis images $\mathbf{r}^s$ and θ^s respectively, and $U(\mathbf{r}^s)$ and $U(\theta^s)$ are the energy functions of the basis images $\mathbf{r}^s$ and θ^s at the sth iteration, respectively.

After r and θ are calculated, basis images x_1 and x_2 can be reconstructed by applying $x_1 = r \cos \theta$ and $x_2 = r \sin \theta$.

2.3   Simulation Setup

To evaluate the proposed methods, a simulated cylindrical phantom was developed. The cross-section related geometric parameters and material composition of the phantom are shown in Fig. 1 and Table 1. Polyethylene (PE) and hydroxyapatite (HA) were used to simulate the soft tissue and bone tissue, and 10% salt water was used to simulate the materials around the tissues. To test the robustness of the proposed material decomposition methods, aluminum (Al), an additional mineral material, was introduced in the phantom, whose mass density was beyond the range of soft tissue and bone tissue [17]. The effective linear attenuation coefficient (ELAC) of each material is given in Table 2.

A fast KV switch dual-energy CT system simulated by Geant4 (Geometry and Tracking) [28] is shown in Fig. 2. A single-slice fan beam was used to scan the phantom, and the fan angle was set to 8.6°. The distances from the source to the

Fig. 1 Cross section of the
cylindrical phantom

Table 1 Geometric parameters and components of the cylindrical phantom in the cross section

Number	Material	Center coordinates (mm)	Radius (mm)	Material composition
I	Aluminum	(−32.5, 0.0)	Inner 4.5, outer 12.5	Al
II	Hydroxyapatite	(30.0, 0.0)	15.0	$Ca_{10}(PO_4)_6(OH)_2$
III	Salt water	(−30.0, 0.0)	15.0	$10\%w(NaCl) + 90\%w(H_2O)$
IV	Polyethylene	(0.0, 0.0)	50.0	$(C_2H_4)_n$
V	Air	(32.5, 0.0)	12.5	$75.5\%w(N_2) + 23.2\%w(O_2) + 1.3\%w(Ar)$

Note: $w(*)$ denotes the mass of each component

Table 2 ELAC of each material under two scanning voltages

	ELAC				
	PE	Salt water	HA	Al	Air
80 kVp	0.2036	0.2859	0.4947	1.100	0.0003
140 kVp	0.1836	0.2340	0.3110	0.7607	0.0002

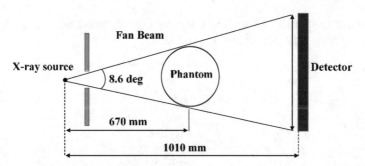

Fig. 2 Diagram of the simulated DECT system

center of the phantom and detector were set to 670 and 1010 mm, respectively, in the simulated X-ray CT system. The incident X-ray energy spectrum was simulated at 80 and 140 kVp with 0.8 mm-thick beryllium and 8.4 mm-thick Al filtration using Spektr software package [29]. The filtration material was used to reduce the radiation damage caused by the low-energy X-ray. The X-ray energy spectra at 80 and 140 kVp are shown in Fig. 3. In the simulation, the rotation angle was set from

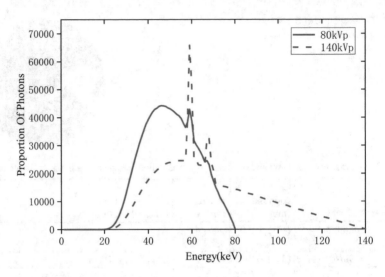

Fig. 3 Normalized energy spectrum at 80 and 140 kVp

0 to 2π, and the angular sampling interval was set to $1°$ with 2×10^6 incident X-ray photons with an energy spectra produced by the above-mentioned method.

2.4 Evaluation Metrics

To evaluate the quality of the basis images quantitatively, noise level (σ) and contrast-to-noise ratio (CNR) are used as follows:

$$\sigma = \sqrt{\frac{1}{N-1} \sum_{n=1}^{N} (z_n - \bar{z})^2} \tag{14}$$

$$\text{CNR} = \frac{|\bar{z}_c - \bar{z}_b|}{\sqrt{\sigma_b^2 + \sigma_c^2}} \tag{15}$$

where σ is the noise level, CNR is the contrast-to-noise ratio, N is the total number of pixels in the region of interest (ROI), z_n is the nth pixel value in the ROI, $\bar{z}$ is the average value of all pixels in the ROI, $\bar{z}_c$ is the average value of pixels in the region of contrast material, σ_c is the noise level of the region of contrast material, $\bar{z}_b$ is the average value of pixels in the region of background material, and σ_b is the noise level of the background material region.

To evaluate the accuracy of the reconstructed image quantitatively, mean square error (MSE) is used as follows:

$$\text{MSE} = \frac{\sum_{n=1}^{N} \left(z_n^{\text{rec}} - z_n^{\text{ref}}\right)^2}{N} \tag{16}$$

where Z_n^{rec} and Z_n^{ref} are the nth pixel values in the reconstructed and reference images, respectively. MSE represents the similarity between the reconstructed image and reference image. For the evaluation of the performance of the proposed basis image reconstruction algorithm, the reference image refers to the theoretical decomposition coefficient image which can be calculated by formula (18) in this paper.

To evaluate the accuracy of material decomposition, the error level of material decomposition is used as follows [30]:

$$\delta = \frac{1}{E_2 - E_1} \int_{E_1}^{E_2} \frac{|\mu(E) - (b_1\mu_1(E) + b_2\mu_2(E))|}{\mu(E)} dE \tag{17}$$

where δ is the error level of material decomposition, and E_1 and E_2 indicate the lowest and highest photon energy to be detected, respectively. Considering the filter material, E_1 is set to 20 keV. E_2 is set to 140 keV, which should not be less than the high voltage used in DECT. $\mu_1(E)$, $\mu_2(E)$, and $\mu(E)$ are the linear attenuation coefficients of the two basis materials and the simulated phantom at energy E.

2.5 Calculation of Theoretical Decomposition Coefficients

Soft and bone tissues were selected as the two basis materials. Hydroxyapatite with a mass density of 0.74 g/cm^3 was the equivalent material of bone tissue and PE with a mass density of 0.97 g/cm^3 was the equivalent material of soft tissue. Theoretical decomposition coefficients were calculated by the effective linear attenuation coefficients of each material at the high and low tube voltages, which can be expressed by Eq. (18). The results are shown in Table 3, in which air is omitted because its effective linear attenuation coefficient is approximately equal to zero.

$$\begin{bmatrix} b_1 \\ b_2 \end{bmatrix} = \begin{bmatrix} \mu_{1L} & \mu_{2L} \\ \mu_{1H} & \mu_{1H} \end{bmatrix}^{-1} \begin{bmatrix} \mu_L \\ \mu_H \end{bmatrix}, \tag{18}$$

where b_1 and b_2 are the decomposition coefficients of soft and bone tissues, L represents the low tube voltage 80 kVp, H represents the high tube voltage 140 kVp, $\mu_{1L/H}$ and $\mu_{2L/H}$ are the effective linear attenuation coefficients of soft and bone tissues at voltages 80 and 140 kVp, respectively, and $\mu_{L/H}$ is the effective linear attenuation coefficients of the phantom at voltages 80 and 140 kVp. The linear attenuation coefficients of the five material insets in the phantom at the energy from 20 to 140 keV are shown in Fig. 4.

Table 3 Soft and bone decomposition coefficients of materials contained in phantom

Material	Mass density (g/cm³)	b_1	b_2
Al	2.69	1.243	1.267
HA	0.74	0	0.740
Salt water	1.07	0.976	0.130
PE	0.97	1.000	0

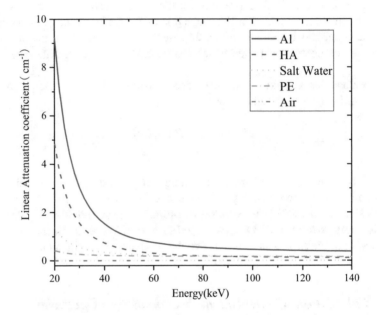

Fig. 4 Linear attenuation coefficients of phantom material [17]

3 Experimental Results

3.1 Evaluation of Regularization Parameter

The term of penalty function in Eqs. (12) and (13) can suppress the noise in the reconstructed basis images. The regularization parameter β controls the strength of the prior relative to the strength of the data in the basis image estimation. A higher β leads to a final estimation which places more emphasis on the prior and less on the data.

To evaluate the effect of parameter β on the accuracy of material decomposition and basis image quality, two metrics are used: (1) the similarity between the reconstructed and the reference images by MSE and (2) the noise level of the basis image by CNR. The smaller the MSE, the more similar is the reconstructed image to the reference image. The larger the CNR, the smaller the noise of image. For the calculation of CNR, PE is selected as the background material in the evaluation.

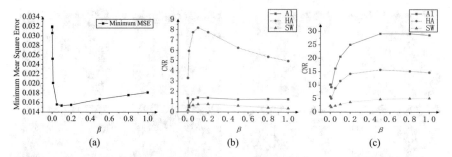

Fig. 5 Minimum MSE and CNRs of Al, HA, and salt water in basis images vs. β with MAP-EM-DD method. (**a**) Minimum MSE. (**b**) CNRs of each material in the basis image of soft tissue. (**c**) CNRs of each material in the basis image of bone tissue

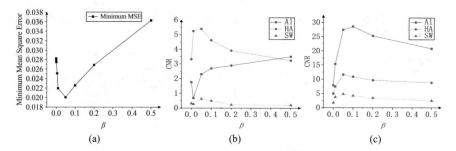

Fig. 6 Minimum MSE and CNRs of Al, HA, and salt water in basis images vs. β with MAP-EM-PT-DD method. (**a**) Minimum MSE. (**b**) CNRs of each material in the basis image of soft tissue. (**c**) CNRs of each material in the basis image of bone tissue

For each parameter β, MSE varies with the change of the iteration number and reaches the minimum at a certain number of iterations. For MAP-EM-DD method, the minimum MSE vs. β and CNRs of each specific material in basis images of soft and bone tissue vs. β are shown in Fig. 5a–c. As shown in Fig. 5a, as the parameter β increases, the minimum MSE decreases first and reaches the minimum at β of approximately 0.1 and then increases gradually. As shown in Fig. 5b, c, the CNRs of Al, HA, and salt water reach the maximum at β of approximately 0.1 in the basis image of soft tissue and 0.5 in the basis image of bone tissue. Considering all the trends of MSE and CNRs vs. β in Fig. 5 together, the parameter β is set to 0.2 in the experiment with MAP-EM-DD method.

With regard to MAP-EM-PT-DD method, the minimum MSE vs. β and CNRs of each specific material in the basis image of soft and bone tissues vs. β are shown in Fig. 6a–c. As shown in Fig. 6a, the minimum MSE reaches the minimum at β of approximately 0.05. As shown in Fig. 6b, c, the CNRs of Al, HA, and salt water reach the maximum at β of approximately 0.05. Therefore, the parameter β is set to 0.05 in the experiment with MAP-EM-PT-DD method.

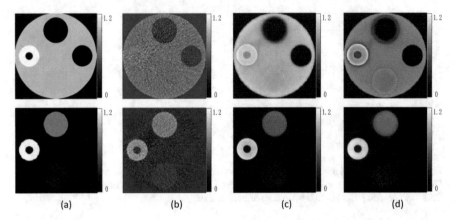

Fig. 7 Basis images reconstructed by (**a**) theoretical calculation, (**b**) FBP-IDD, (**c**) MAP-EM-DD, and (**d**) MAP-EM-PT-DD

Table 4 Noise levels of basis images

	Noise level (σ)		
	FBP-IDD	MAP-EM-DD	MAP-EM-PT-DD
Soft tissue	0.209	0.089	0.068
Bone tissue	0.077	0.028	0.022

3.2   Comparison of the Basis Images

To reconstruct the basis images with MAP-EM-DD and MAP-EM-PT-DD methods, the iteration stop condition is set as the difference of MSE between two consecutive reconstructed images to less than 1e−6. The theoretical basis images calculated by the average attenuation coefficients of each specific material in the phantom are shown in Fig. 7a. The basis images reconstructed by FBP-IDD, MAP-EM-DD, and MAP-EM-PT-DD methods are shown in Fig. 7b–d, respectively. The first row in Fig. 7 shows the basis images of soft tissue and the second row shows the bone tissue.

3.2.1   Comparison of the Noise Levels (σ)

The noise levels (σ) of basis images reconstructed by FBP-IDD, MAP-EM-DD, and MAP-EM-PT-DD methods are calculated and compared. The noise levels (σ) of basis images are shown in Table 4. Compared with FBP-IDD method, MAP-EM-DD method can reduce the noise levels (σ) of the basis images of soft tissue and bone tissue by 57.4% and 63.6%, respectively. Furthermore, compared with MAP-EM-DD method, MAP-EM-PT-DD method can reduce the noise levels (σ) of the basis images of soft tissue and bone tissue by 23.6% and 21.4%, respectively.

3.2.2   Comparison of the Gray Level Distributions in the Decomposition Coefficient Images

To compare the gray level distributions in the decomposition coefficient images reconstructed by the methods of FBP-IDD, MAP-EM-DD, and MAP-EM-PT-DD with theoretical gray level distribution, the red lines with arrows are plotted in the decomposition coefficient images with the above methods respectively, as illustrated in Fig. 8a, c. The gray level distribution curves through the red lines are shown in Fig. 8b, d. Figure 8b shows the curves of gray level distribution along the red line with arrows located in the theoretical basis image of soft tissue and the decomposition coefficient images of soft tissue reconstructed by the methods of FBP-IDD, MAP-EM-DD, and MAP-EM-PT-DD, respectively. Figure 8d shows the curves of gray level distribution along the red line with arrows located in the theoretical basis image of bone tissue and the decomposition coefficient image of bone tissue reconstructed by the methods of FBP-IDD, MAP-EM-DD, and MAP-EM-PT-DD, respectively.

As shown from the curves of gray level distribution in Fig. 8b, d, we can see that the curves of gray distribution along the red lines in the decomposed images of soft tissue reconstructed by FBP-IDD method have the highest fluctuation and they are quite different from the curves of theoretical gray distribution. The fluctuation of the curves of gray level distribution in the decomposed images reconstructed by MAP-EM-DD and MAP-EM-PT-DD methods is much smaller than that reconstructed by FBP-IDD. The difference between the curves of gray level distribution in the decomposed images reconstructed by MAP-EM-DD and MAP-EM-PT-DD methods is small, the trends of the curves are similar, and the curves are basically consistent to the curves of theoretical gray distribution; however, the range of fluctuation of the curves of gray level distribution in the decomposition coefficient image of bone tissue is less than that of the curves of gray level distribution in the decomposition coefficient image of soft tissue. It implies that the noise level of the decomposition coefficient image of soft tissue is higher than that of the decomposition coefficient image of bone tissue. The accuracy of the decomposition coefficient images reconstructed by the direct iterative material decomposition methods, i.e., MAP-EM-DD and MAP-EM-PT-DD methods, is higher than that of the image domain material decomposition method, i.e., FBP-IDD method, and the noise level of the decomposition coefficient images reconstructed by the direct iterative material decomposition method, i.e., MAP-EM-DD and MAP-EM-PT-DD methods, is lower than that of the image domain material decomposition method, i.e., FBP-IDD method.

3.2.3   Comparison of Contrast Noise Ratios

Table 5 shows the CNRs of material-specific regions in basis images, where CNR_1, CNR_2, and CNR_3 represent the CNR of basis images reconstructed by the methods of FBP-IDD, MAP-EM-DD, and MAP-EM-PT-DD, respectively. As shown in Table

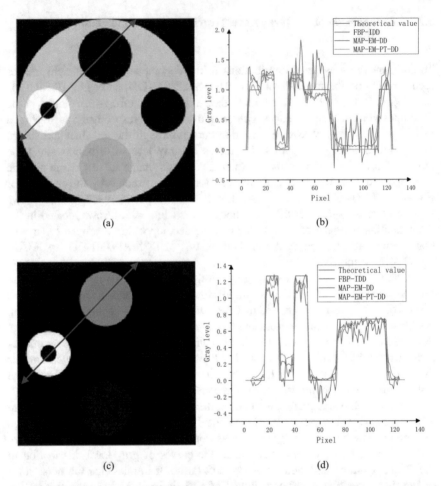

Fig. 8 Illustration of basis images of soft and bone tissues and the corresponding curves of gray level distribution along the red line with arrows located in the theoretical basis images and the basis images reconstructed by the methods of FBP-IDD, MAP-EM-DD, MAP-EM-PT-DD. (**a**) Basis image of soft tissue. (**b**) The curves of gray level distribution along the red lines located in the basis images of soft tissue. (**c**) Basis image of bone tissue. (**d**) The curves of gray level distribution along the red line located in the basis images of bone tissue

5, $CNR_2 > CNR_1$ and $CNR_3 > CNR_1$ in all the material-specific regions in basis images. Compared with FBP-IDD method, MAP-EM-DD method can improve the CNRs ranging from 63.8% to 237.3%. Furthermore, $CNR_3 > CNR_2$ in the material-specific regions of Al in basis images, whereas $CNR_3 < CNR_2$ in the material-specific regions of HA in basis images, and CNR_2 and CNR_3 are almost equivalent in the region of salt water in basis images. Compared with MAP-EM-DD method, MAP-EM-PT-DD method can improve the CNRs ranging from 9.4% to 68.5% in the material-specific regions of Al in basis images and by 30.0% in the region of salt

Table 5 CNRs of material-specific regions in basis images

CNR									
	Al			HA			Salt water		
	CNR_1	CNR_2	CNR_3	CNR_1	CNR_2	CNR_3	CNR_1	CNR_2	CNR_3
Soft tissue	0.832	1.363	2.297	3.229	7.754	5.386	0.301	0.762	0.622
Bone tissue	7.419	25.022	27.375	5.327	14.170	11.604	1.240	3.694	4.802

Table 6 Average decomposition coefficients of the specific materials in the basis images

	Al		HA		Salt water		PE	
	b_1	b_2	b_1	b_2	b_1	b_2	b_1	b_2
FBP-IDD	1.382	1.011	0.098	0.634	1.003	0.147	1.092	0.014
MAP-EM-DD	1.194	1.221	0.111	0.686	0.985	0.137	1.020	0.018
MAP-EM-PT-DD	1.259	1.197	0.088	0.682	0.980	0.140	1.041	0.004

water in the basis image of bone tissue and reduce the CNR by 18.4% in the region of salt water in the basis image of soft tissue, and reduce the CNRs ranging from 18.1% to 30.5% in the material-specific regions of HA in basis images.

3.2.4 Comparison of Error Levels of the Decomposition Coefficients

Table 6 and Fig. 9 show the averages and error levels of the decomposition (δ) coefficients of the specific materials by the methods of FBP-IDD, MAP-EM-DD, and MAP-EM-PT-DD. Compared with FBP-IDD method, MAP-EM-DD method can reduce the error levels of decomposition (δ) coefficients of HA, PE, salt water, and Al by 31.7%, 45.7%, 45.8%, and 62.1%, respectively. Compared with MAP-EM-DD method, MAP-EM-PT-DD method can reduce the error levels of decomposition (δ) coefficients of salt water, Al, PE, and HA by 1.9%, 17.9%, 26.9%, and 36.3%, respectively.

3.2.5 Comparison of the Reconstruction Time of Basis Images

The cost time for the basis image reconstruction is also critical in CT applications. In general, FBP algorithm is faster than iterative algorithms. To compare the computation performance of MAP-EM-DD and MAP-EM-PT-DD methods quantitatively, their initialization of the basis images and stop condition of iteration are set to be the same. The reconstruction of basis images is implemented on a laptop with 2.6GHz Intel Core i7-8850H CPU and 16GB RAM using Matlab 2017a (Mathworks, Inc.). The reconstruction times of basis images with the methods of FBP-IDD, MAP-EM-DD, and MAP-EM-PT-DD are 48 s, 718 s, and 617 s, respectively. FBP-IDD method takes the least reconstruction time. Compared with MAP-EM-DD method, MAP-EM-PT-DD method can reduce the reconstruction time by 14.1%.

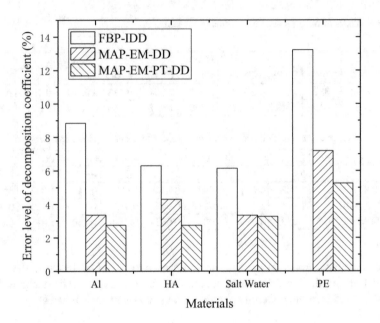

Fig. 9 Comparison of error levels of decomposition (δ) coefficients of the specific materials

Therefore, FBP-IDD method can be employed when quick decomposition is considered but not high precision, and MAP-EM-PT-DD method is the most suitable method when both accuracy and time are considered. Furthermore, it is encouraged to perform the iterative procedure of both MAP-EM-DD and MAP-EM-PT-DD methods with GPU for a quick and accurate basis image reconstruction.

4   Discussion

Spectral CT can provide information on material characterization and quantification by material decomposition algorithms. However, traditional methods suffer from the non-ideal effects of X-ray imaging systems, such as the noise and spectrum distortion in measurements, resulting in the decreased accuracy of material decomposition.

Our previous studies have shown that MAP-EM algorithm can improve the quality of spectral CT image effectively. Based on this finding, we proposed a direct iterative material decomposition method based on MAP-EM algorithm (MAP-EM-DD) and an alternative method (MAP-EM-PT-DD) to improve the accuracy of material decomposition in this work. The performance of the proposed methods was evaluated and compared with FBP-IDD method using an established cylinder phantom. The effect of the regularization parameter β on the accuracy of material decomposition and quality of basis images were investigated. The metrics of noise level, CNR, and error level were calculated and compared to evaluate the performance of the proposed methods.

One-step methods perform decomposition and reconstruction simultaneously. Unlike two-step methods, one-step methods can circumvent the drawbacks of information loss, improve image quality, and reduce noise-induced bias. Considering the advantages of one-step methods and MAP-EM algorithm, MAP-EM-DD method should be able to improve the accuracy of material decomposition for spectral CT. Incorporating the polar coordinate transformation into MAP-EM-DD, MAP-EM-PT-DD method is proposed. The transformation can make the basis images more relevant during iteration, which can benefit material decomposition and basis image reconstruction.

The results demonstrate that MAP-EM-DD method can remarkably reduce the noise in the decomposed basis images and the error of material decomposition and improve the CNR of the basis images. This result is because the statistical properties of the projection and preservation of information are considered in MAP-EM-DD method, imposing the prior knowledge of the basis images on the material decomposition to suppress noise. Furthermore, compared with MAP-EM-DD method, MAP-EM-PT-DD method performs better with regard to noise level, error level, and reconstruction time, while comparable with respect to CNR. The latter method saves more than 14% of reconstruction time. This result may be contributed to the more relevant of basis images during the iteration procedure using r and θ in the polar coordinate system. However, the edge-gradient effect for MAP-EM-PT-DD method is worse than that for MAP-EM-DD method.

Although MAP-EM-DD and MAP-EM-PT-DD methods can remarkably improve the accuracy of material decomposition and basis image quality, this study bears certain limitations. Firstly, the proposed methods should be compared with other one-step methods for further performance evaluation. Secondly, the performance of the proposed methods with projection datasets from commercial spectral CT system should be validated. Thirdly, it is necessary to investigate how to reduce the edge-gradient effect in the proposed algorithms as that in the regular CT image reconstruction. Finally, the improvement of computation speed of the proposed methods is encouraged for practical use.

5   Conclusion

Two novel and robust one-step material decomposition methods, MAP-EM-DD and MAP-EM-PT-DD, are proposed in this paper. MAP-EM-DD method outperforms FBP-IDD method with regard to noise level, CNR, and error level. MAP-EM-PT-DD method outperforms MAP-EM-DD method with regard to noise level, error level, and reconstruction time. The proposed methods can evidently improve the accuracy of material decomposition and improve the quality of basis images for spectral CT. The suitable regulation parameters are also necessary to achieve high accuracy of material decomposition and high quality of basis images.

Acknowledgments I wish to thank all my coworkers in the State Key Laboratory of Mechanics and Control of Mechanical Structures and the Department of Nuclear Science and Engineering at Nanjing University of Aeronautics and Astronautics, especially my Master's students, Runchao Xin and Xuling Zhang, for their continuous research efforts on this exciting new X-ray imaging science and technology. I also thank Dr. Iniewski for the invitation and the support in part by the National Natural Science Foundation of China (51575256), Key Research and Development Plan (Social Development) of Jiangsu Province (BE2017730), Key Industrial Research and Development Plan of Chongqing (cstc2017zdcy-zdzxX0007), Shanghai Aerospace Science and Technology Innovation Fund (SAST 2019-121), and the Priority Academic Program Development of Jiangsu Higher Education Institutions (PAPD).

References

1. Alvarez RE, Macovski A. Energy-selective reconstructions in X-ray computerized tomography. Phys Med Biol. 1976;21(5):733–44. https://doi.org/10.1088/0031-9155/21/5/002.
2. Sidky EY, Zou Y, Pan X. Impact of polychromatic x-ray sources on helical, cone-beam computed tomography and dual-energy methods. Phys Med Biol. 2004;49(11):2293–303. https://doi.org/10.1088/0031-9155/49/11/012.
3. Kim C, Kim D, Cha J, Kim H, Choo JY, Cho P-K, Lee KY. The optimal energy level of virtual monochromatic images from spectral CT for reducing beam-hardening artifacts due to contrast media in the thorax. Am J Roentgenol. 2018;211(3):557–63. https://doi.org/10.2214/AJR.17.19377.
4. Schmidt TG, Sammut BA, Barber RF, Pan X, Sidky EY. Addressing CT metal artifacts using photon-counting detectors and one-step spectral CT image reconstruction. Med Phys. 2022;49(5):3021–40. https://doi.org/10.1002/mp.15621.
5. Thieme SF, Graute V, Nikolaou K, Maxien D, Johnson TRC. Dual energy CT lung perfusion imaging–correlation with SPECT/CT. Eur J Radiol. 2010;81(2):360–5. https://doi.org/10.1016/j.ejrad.2010.11.037.
6. Busi M, Mohan KA, Dooraghi AA, Champley KM, Martz HE, Olsen UL. Method for system-independent material characterization from spectral X-ray CT. NDT&E Int. 2019;107:102136. https://doi.org/10.1016/j.ndteint.2019.102136.
7. Lambert JW, Sun Y, Gould RG, Ohliger MA, Li Z, Yeh BM. An image-domain contrast material extraction method for dual-energy computed tomography. Investig Radiol. 2017;52(4):245–54. https://doi.org/10.1097/RLI.0000000000000335.
8. Xie B, Su T, Kaftandjian V, Niu P, Yang F, Robini M, Zhu Y, Duvauchelle P. Material decomposition in X-ray spectral CT using multiple constraints in image domain. J Nondestruct Eval. 2019;38(1):16. https://doi.org/10.1007/s10921-018-0551-8.
9. Maass C, Baer M, Kachelriess M. Image-based dual energy CT using optimized precorrection functions: a practical new approach of material decomposition in image domain. Med Phys. 2009;36(8):3818–29. https://doi.org/10.1118/1.3157235.
10. Wu WW, Yu HJ, Chen PJ, Luo FL, Liu FL, Wang Q, Zhu YN, Zhang YB, Feng J, Yu HY. Dictionary learning based image-domain material decomposition for spectral CT. Phys Med Biol. 2020;65(24):245006. https://doi.org/10.1088/1361-6560/aba7ce.
11. Schlomka JP, Roessl E, Dorscheid R, Dill S, Martens G, Istel T, Bäumer C, Herrmann C, Steadman R, Zeitler G, Livne A, Proksa R. Experimental feasibility of multi-energy photon-counting K-edge imaging in pre-clinical computed tomography. Phys Med Biol. 2008;53(15):4031–47. https://doi.org/10.1088/0031-9155/53/15/002.
12. Sawatzky A, Xu Q, Schirra CO, Anastasio MA. Proximal ADMM for multi-channel image reconstruction in spectral x-ray CT. IEEE Trans Med Imaging. 2014;33(8):1657–68. https://doi.org/10.1109/tmi.2014.2321098.

13. Ducros N, Abascal JFPJ, Sixou B, Rit S, Peyrin F. Regularization of nonlinear decomposition of spectral x-ray projection images. Med Phys. 2017;44(9):e174–87. https://doi.org/10.1002/mp.12283.

14. Xie HQ, Ren Y, Long WT, Yang XF, Tang XY. Principal component analysis in projection and image domains another form of spectral imaging in photon-counting CT. IEEE Trans Biomed Eng. 2021;68(3):1074–83. https://doi.org/10.1109/TBME.2020.3013491.

15. Mory C, Sixou B, Si-Mohamed S, Boussel L, Rit S. Comparison of five one-step reconstruction algorithms for spectral CT. Phys Med Biol. 2018;63:235001. https://doi.org/10.1088/1361-6560/aaeaf2.

16. Foygel BR, Sidky EY, Gilat ST, Pan X. An algorithm for constrained one-step inversion of spectral CT data. Phys Med Biol. 2016;61(10):3784–818. https://doi.org/10.1088/0031-9155/61/10/3784.

17. Cai C, Rodet T, Legoupil S, Mohammad-Djafari A. A full-spectral Bayesian reconstruction approach based on the material decomposition model applied in dual-energy computed tomography. Med Phys. 2013;40(11):111916. https://doi.org/10.1118/1.4820478.

18. Long Y, Fessler JA. Multi-material decomposition using statistical image reconstruction for spectral CT. IEEE Trans Med Imaging. 2014;33(8):1614–26. https://doi.org/10.1109/TMI.2014.2320284.

19. Weidinger T, Buzug TM, Flohr T, Kappler S, Stierstorfer K. Polychromatic iterative statistical material image reconstruction for photon-counting computed tomography. Int J Biomed Imaging. 2016;2016:5871604. https://doi.org/10.1155/2016/5871604.

20. Mechlem K, Ehn S, Sellerer T, Braig E, Munzel D, Pfeiffer F, Noël PB. Joint statistical iterative material image reconstruction for spectral computed tomography using a semi-empirical forward model. IEEE Trans Med Imaging. 2018;37(1):68–80. https://doi.org/10.1109/TMI.2017.2726687.

21. Fang W, Wu DF, Kim K, Kalra MK, Singh R, Li L, Li QZ. Iterative material decomposition for spectral CT using self-supervised Noise2Noise prior. Phys Med Biol. 2021;66(15):155013. https://doi.org/10.1088/1361-6560/ac0afd.

22. Zhou ZD, Xin RC, Guan SL, Li JB, Tu JL. Investigation of maximum a posteriori probability expectation-maximization for image-based weighting spectral X-ray CT image reconstruction. J Xray Sci Technol. 2018;26(5):853–64. https://doi.org/10.3233/XST-180396.

23. Zhou ZD, Guan SL, Xin RC, Li JB. Investigation of contrast-enhanced subtracted breast CT images with MAP-EM based on projection-based weighting imaging. Australas Phys Eng Sci Med. 2018;41:371–7. https://doi.org/10.1007/s13246-018-0634-y.

24. Maaß C, Meyer E, Kachelrieß M. Exact dual energy material decomposition from inconsistent rays (MDIR). Med Phys. 2011;38(2):691–700. https://doi.org/10.1118/1.3533686.

25. Chen Y, Ma JH, Feng QJ, Luo LM, Shi PC, Chen WF. Nonlocal prior Bayesian tomographic reconstruction. J Math Imaging Vis. 2008;30(2):133–46. https://doi.org/10.1007/s10851-007-0042-5.

26. Green PJ. Bayesian reconstructions from emission tomography data using a modified EM algorithm. IEEE Trans Med Imaging. 1990;9(1):84–93. https://doi.org/10.1109/42.52985.

27. Brooks RA. A quantitative theory of the Hounsfield unit and its application to dual energy scanning. J Comput Assist Tomogr. 1977;1(4):487–93. https://doi.org/10.1097/00004728-197710000-00016.

28. Agostinelli S, Allison J, Amako K, et al. Geant4—a simulation toolkit. Nucl Instrum Methods Phys Res A. 2003;506(3):250. https://doi.org/10.1016/S0168-9002(03)01368-8.

29. Punnoose J, Xu J, Sisniega A, Zbijewski W, Siewerdsen JH. Technical note: spektr 3.0-a computational tool for x-ray spectrum modeling and analysis. Med Phys. 2016;43(8):4711–7. https://doi.org/10.1118/1.4955438.

30. Zhang GW, Cheng JP, Zhang L, Chen ZQ, Xing YX. A practical reconstruction method for dual energy computed tomography. J Xray Sci Technol. 2008;16(2):67–88.

Quantitative Analysis Methodology of X-Ray Attenuation for Medical Diagnostic Imaging: Algorithm to Derive Effective Atomic Number, Soft Tissue and Bone Images

Natsumi Kimoto, Hiroaki Hayashi, Cheonghae Lee, Tatsuya Maeda, Daiki Kobayashi, Rina Nishigami, and Akitoshi Katsumata

1   Introduction

When X-rays were discovered by Dr. W. Röntgen in 1895, the publication of an X-ray image of the human hand had a tremendous impact on the progress of medicine. In his first paper on the discovery of X-rays, he published experimentally derived properties of X-rays and X-ray photographs. Although he described abstract trends related to the attenuation of X-rays, much more time was needed to elucidate this physical phenomenon based on discoveries, such as the Photoelectric and Compton scattering effects, and the maturity of corresponding quantum theories. Fortunately, the development of medicine did not necessarily require the physics theory related to X-ray attenuation. This is because medical doctors were able to understand radiographs without having knowledge of the physics theory behind X-rays. At that time, X-ray image diagnosis was performed by a doctor who was looking at an X-ray image reflected on a fluorescent plate. Gradually, this came to be replaced by radiographs. When reviewing the history of X-ray photography, we should pay tribute to the historical success of radiography in qualitative diagnosis, and at the same time, we should notice the fact that the theoretical consideration of X-rays had not been reflected in the diagnosis using X-ray photographs.

N. Kimoto (✉)
College of Medical, Pharmaceutical and Health Sciences, Kanazawa University, Ishikawa, Japan

Job Corporation, Kanagawa, Japan

H. Hayashi
College of Medical, Pharmaceutical and Health Sciences, Kanazawa University, Ishikawa, Japan

C. Lee · T. Maeda · D. Kobayashi · R. Nishigami
Graduate School of Medical Sciences, Kanazawa University, Ishikawa, Japan

A. Katsumata
Department of Oral Radiology, Asahi University, Gifu, Japan

© The Author(s), under exclusive license to Springer Nature Switzerland AG 2023
S. Hsieh, K. (Kris) Iniewski (eds.), *Photon Counting Computed Tomography*,
https://doi.org/10.1007/978-3-031-26062-9_11

Since then, X-ray images have become increasingly essential for medical examination and non-invasive inspection. In particular, the digitization of X-ray images using various image processing technologies, such as computed radiography (CR) and digital radiography (DR) systems, has made it possible to generate X-ray images that are easier to use for quantitative diagnoses [1, 2]. In recent years, new attempts to incorporate artificial intelligence into diagnostic technology are also attracting attention [3]. Although these techniques focus on how much useful information can be extracted from a generated X-ray image, it should be pointed out that the amount of information which is derived from the original X-ray image is limited. If the detection and analysis technologies of X-rays are improved, we would be able to derive more information from X-ray images. We would like to develop a new image generation technique and investigate new possibilities for X-ray images.

The concept of the diagnostic system that we want to propose is fusion of conventional and novel methods. In conventional diagnosis, a medical doctor interprets a gray-scale image reflecting the absorbed X-ray energy based on anatomical knowledge. It means that a biological perspective plays an important role in this diagnosis. On the other hand, our goal is to carefully analyze the interaction of X-rays related to image generation; during the production of X-ray images, X-rays penetrate the human body and further interact with detector materials. The analysis of the interaction in each process will lead to the generation of a novel concept X-ray image based on the perspective of physics. We expect that the fusion of conventional qualitative diagnosis based on biology and our proposed quantitative diagnosis based on physics will lead to innovations in diagnostic performance.

Currently, an energy-resolving photon-counting detector (ERPCD) which can generate images corresponding to the energy of each X-ray is being developed [4–6]. In our research, we focus on two-dimensional X-ray images and devise ways to make the best use of the X-ray attenuation information obtained from the ERPCD. In this chapter, we explain a novel method to identify materials using an ERPCD. First, we will explain the basic principle of material identification based on analysis of X-ray attenuation. Next, we will describe the issues which need to be considered when analyzing X-rays detected by ERPCD. Finally, we will present algorithms to derive an effective atomic number (Z_{eff}) image and to extract mass thickness (ρt) images related to soft tissue and bone. Although the main topic of this chapter is to explain the novel method based on analysis of X-ray attenuation, the physics of X-ray attenuation is common for generating X-ray images using various systems such as plain X-ray and computed tomography (CT) examinations, and we are confident that this chapter will be useful for future research using ERPCDs. If readers are interested in these contents, we recommend reading our papers [7–16] and books [17–19] for more detailed information.

2 Energy-Resolving Photon-Counting Detector to Realize Quantitative Analysis of an Object

Next-generation-type ERPCD has been developed to realize novel analysis for such things as extracting properties of an object [4–6]. This section describes the processing procedure of the ERPCD, and the principle of material identification based on the analysis of X-ray attenuation.

To clearly understand the mechanism of an ERPCD, we need to compare it with an energy integrating detector (EID) [20, 21] which has been traditionally applied to medical and industrial X-ray examinations, as shown in Fig. 1. Figure 1a shows a schematic drawing of an EID. Assuming that when four X-rays having different energies of E_1, E_2, E_3, and E_1 are incident to the pixel of the EID, corresponding charge clouds are generated in a pixel. Then, they are collected by an integrated circuit, and the corresponding totally absorbed energy ($\sum_i E_i$) values are digitalized and presented in a gray-scale image; this is the processing procedure for a conventional X-ray image. On the other hand, an ERPCD can analyze X-rays individually and discriminate X-ray energies corresponding to several energy thresholds [4–6]. Figure 1b shows a schematic drawing of an ERPCD. When setting threshold levels at $E_0{}'$, $E_1{}'$, $E_2{}'$, and $E_3{}'$, we can separate signals into three energy regions of $E_0{}' < E_i < E_1{}'$, $E_1{}' < E_i < E_2{}'$, $E_2{}' < E_i < E_3{}'$. In the case in which four X-rays having energies of E_1, E_2, E_3, and E_1 are incident to the pixel of ERPCD, E_1, E_2, and E_3 are classified into low-, middle-, and high-energy bins, respectively. This is a main difference from the EID, which is used to measure not the energy of an X-ray but the number of X-rays. In this case, two counts are measured in the low-energy bin and one count is measured in each of the middle- and high-energy bins. We would like to create quantitative images that extract the properties of an object based on analysis using X-ray energy information. We can also produce a conventional X-ray image which provides qualitative information because the image related to the energy bin has information concerning intensity and effective energy.

Image generation using an ERPCD can quantitatively analyze the X-ray attenuation of an object. Here, we will introduce the method to extract the properties of an object from X-ray energy information. Assuming that an object having a certain Z_{eff} is measured, X-ray attenuation can be determined using the well-known formula [22]:

$$I = I_0 \times \exp(-\mu t), \tag{1}$$

where I_0 and I are the intensities of the incident and penetrating X-rays, respectively, and μ and t are the linear attenuation coefficient and material thickness, respectively. It is important to understand the fact that the probability of X-ray attenuation, μ, is determined by the atomic number and a function of the energy of the X-rays [22]. This means that if we can derive μ corresponding to a certain X-ray energy, we can determine the atomic number of an object. However, the physical quantity obtained from one piece of energy information is μt, and we cannot derive μ without using

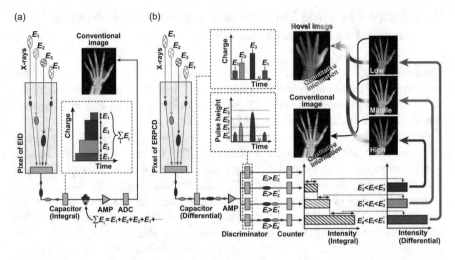

Fig. 1 Comparison between (**a**) energy integrating detector (EID) and (**b**) energy-resolving photon-counting detector (ERPCD). The EID generates images with information related to the sum of the energies. On the other hand, the ERPCD which is the focus of our research can analyze each X-ray energy

the information for t. To extract μ from μt, dual energy information is required and ERPCD can be used to provide information corresponding to multiple energies with a single irradiation. In this study, we focus our attention on the low-energy E_{low} and high-energy E_{high} and calculate the corresponding μt values as $\mu_{\text{low}} t$ and $\mu_{\text{high}} t$. Then, taking the ratio of these values, we can derive the value $\frac{\mu_{\text{low}}}{\mu_{\text{high}}}$ $(= \frac{\mu_{\text{low}} t}{\mu_{\text{high}} t})$ which can be uniquely determined by atomic number without information for t [23]. In order to determine the Z_{eff} from $\frac{\mu_{\text{low}}}{\mu_{\text{high}}}$, a unique relationship between $\frac{\mu_{\text{low}}}{\mu_{\text{high}}}$ and Z_{eff} is available, and these data can be derived from the reference data μ/ρ [24]. When using the theoretical relationship, experimentally derived $\frac{\mu_{\text{low}}}{\mu_{\text{high}}}$ can be converted into Z_{eff}. This means that a novel image that reflects the Z_{eff} of an object can be produced by analyzing X-ray energy information generated from an ERPCD. We should note that this description is based on ideal energy absorption resulting from the interaction between incident X-rays and detector materials and focuses on monochromatic X-ray energy. However, actual cases are far from the ideal condition and, therefore, the value measured with an ERPCD needs to be corrected to a value related to monochromatic X-ray.

3 Issues That Need to Be Considered to Achieve Accurate Material Identification

In the description in the former section, we explained that X-ray energy information needs to be analyzed accurately in order to derive the properties of an object. In general, medical, and industrial applications, (1) polychromatic X-rays and (2) multi-pixel-type imaging detectors are widely applied and, therefore, it is difficult to handle X-ray energy information ideally. In order to clarify the issues caused by these two restrictions, we show the differences in X-ray spectra obtained in each stage from incident to detection as shown in Fig. 2. The upper right panel shows the X-ray spectrum in which X-rays having various energies are generated from the X-ray equipment. When using an ERPCD having two energy bins, the low- and high-energy bins, presented by red and blue respectively, are analyzed quantitatively. To perform the analysis, it is necessary to distinguish the energies, and we generally use the effective energies of each energy bin. In the spectrum obtained after penetrating an object, the degree of X-ray attenuation varies in the difference of Z_{eff} of an object, and as a result, the effective energy changes. This phenomenon is known as the "beam hardening effect" [25]. Furthermore, when X-rays are incident to the detector, the X-rays interact with detector materials, and the corresponding X-ray spectrum generated from the detector is much different than the incident X-ray spectrum. Although current commercial products use to take a practical approach for material decomposition based on the actual X-ray spectrum [26–29], our approach is much different. We apply the correction for the beam hardening effect and detector response to actual data by taking into consideration the Z_{eff} of an object [14–16, 19]. This method can process actual polychromatic X-rays as those equivalent to ideal monochromatic X-rays, which results in the accomplishment of what is considered "ideal" analysis based on monochromatic X-rays. In the following sections, we will carefully analyze the effects of the beam hardening effect and detector response and also introduce a novel method to correct these effects.

3.1 Beam Hardening Effect Depends on Effective Atomic Number

In this section, we will explain the beam hardening effect which is one of the issues to be considered for precise material identification. Figure 3 shows the relationship between the beam hardening effect and Z_{eff}. Figure 3a shows Z_{eff} values of the elements which compose the human body. The human body targeted for clinical examination is composed of various organs and tissue, and the Z_{eff} values of each organ and tissue are distributed at relatively low atomic number values. These Z_{eff}

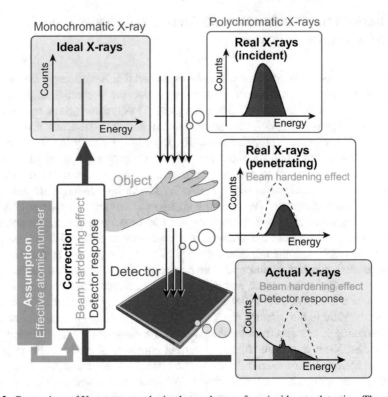

Fig. 2 Comparison of X-ray spectra obtained at each stage from incident to detection. The actual X-ray spectrum includes both the beam hardening effect and detector response. Our approach is to treat the measured polychromatic X-rays as if they are monochromatic X-rays by correcting for the beam hardening effect and detector response

values can be calculated from the elemental composition of each organ [24] using a well-known formula [30]. As shown in this figure, adipose tissue is 6.33, muscle is 7.45, and cortical bone is 13.23; it is obvious that there is a slight difference in Z_{eff} between the soft tissues, and it can be roughly divided into soft tissue at around $Z_{eff} = 7$ and bone around $Z_{eff} = 13$. Next, we present the beam hardening effect depending on the difference of Z_{eff}. Figure 3b shows the changes in the spectra and the effective energies ($\overline{E}$) corresponding to objects having various Z_{eff} values. The black line is the incident X-ray spectrum at 50 kV which is reproduced by a semi-empirical formula [31], and the $\overline{E}$ of the energy region 20–32 keV is 27.4 keV. We also present the X-ray spectra after penetrating objects of which Z_{eff} values and ρt value are set to be 5–15 and 1 g/cm^2, respectively. The distribution of X-rays is different depending on Z_{eff}, and it is important to especially focus on $Z_{eff} = 7$ (blue line) and $Z_{eff} = 13$ (red line). The corresponding $\overline{E}$ values are 27.8 keV and 29.5 keV, respectively. It is found that the difference between incident X-ray and penetrating X-ray shows the beam hardening effect, and the degree of effect differs depending on the Z_{eff}. Unless these effects are properly corrected based

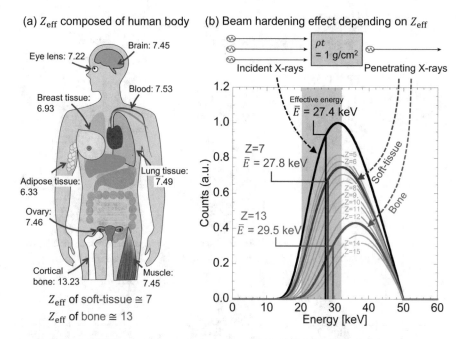

Fig. 3 Relationship between beam hardening effect and Z_{eff}. (**a**) The Z_{eff} values of various tissues which compose the human body. (**b**) The changes in the X-ray spectra and effective energies corresponding to objects having various Z_{eff} values

on the information of an object, it is not possible to perform an analysis in which polychromatic X-rays are regarded as monochromatic X-rays.

3.2 Response of Multi-Pixel-Type Energy-Resolving Photon-Counting Detector

Another issue to be considered for precise material identification is detector response. In this section, we will explain the concept of the response function of a multi-pixel-type ERPCD and the procedure to reproduce an X-ray spectrum using the response function. Figure 4 shows a concept of the various phenomena related to detector response when X-rays Φ are incident to the detector. Figure 4a shows an ideal case where all of the energies of the incident X-rays are absorbed completely by the detector. Using the response function represented by an identity matrix $\mathbf{I}$, the corresponding X-ray spectrum can be calculated by the matrix operation $\mathbf{I\Phi}$ ($= \Phi$). In actuality, the response function including phenomenon $\mathbf{R}^{(1)}$, which is related to radiation interactions, and phenomenon $\mathbf{R}^{(2)}$, which is related to electric charge collecting processes, should be considered to reproduce an X-ray spectrum measured with a multi-pixel-type ERPCD. Figure 4b-1 shows $\mathbf{R}^{(1)}$ in which we

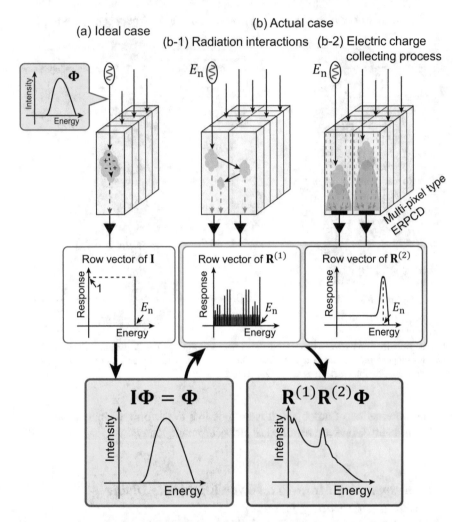

Fig. 4 Comparison between detector responses for ideal and actual cases. (**a**) In the ideal case, incident X-rays Φ are totally absorbed within the pixel of the detector, and the detected X-ray spectrum ($\mathbf{I}\Phi = \Phi$) is the same as the incident X-ray spectrum. (**b**) In actuality, incident X-rays are affected by two different phenomena: (**b-1**) radiation interactions and (**b-2**) electric charge collecting processes. Consequently, the detected X-ray spectrum $\mathbf{R}^{(1)}\mathbf{R}^{(2)}\Phi$ is much different from that of the incident X-ray spectrum

take into consideration the transportation of the characteristics X-rays caused by the photoelectric effect and scattered X-rays due to the Compton scattering effect [7, 9, 17, 19, 32–36]. Figure 4b-2 shows $\mathbf{R}^{(2)}$ in which the charge sharing effect [32–36] and energy resolution [32, 33, 35, 36] are considered. When considering both $\mathbf{R}^{(1)}$ and $\mathbf{R}^{(2)}$ operation of Φ, the X-ray spectrum under actual conditions can then be reproduced by matrix operation $\mathbf{R}^{(1)}\mathbf{R}^{(2)}\Phi$. This model can predict the accurate detector response under the assumption in which a contamination of scattered X-rays and a generation of pulse pile-up are not presented.

Here, we will look more closely at the response functions $\mathbf{R}^{(1)}$ and $\mathbf{R}^{(2)}$. First, the response function $\mathbf{R}^{(1)}$ is calculated by simulating the interaction between the incident X-rays and detector materials using Monte-Carlo simulation code EGS5 [37]. The detector materials are composed of cadmium zinc telluride (CZT) at a ratio of Cd:Zn:Te = 0.9:0.1:1.0 with a density of 5.8 g/cm^3 and the outer size of the monolithic detector material is set as 10 mm × 10 mm with a thickness of 1.5 mm. The response function is defined as the spectra related to a center pixel of 200 μm × 200 μm. The 0–140 keV monochromatic X-rays of 10^6 photons are irradiated to the center pixel, and the total irradiation area is determined so as to establish equilibrium for the secondary produced radiation [9, 17, 19]. A two-dimensional matrix $\mathbf{R}^{(1)}$ is constructed from the elements of response functions corresponding to 0–140 keV monochromatic X-rays. Here, we demonstrate 80 keV monochromatic X-ray vector element for $\mathbf{R}^{(1)}$ as presented in Fig. 5a. The red line

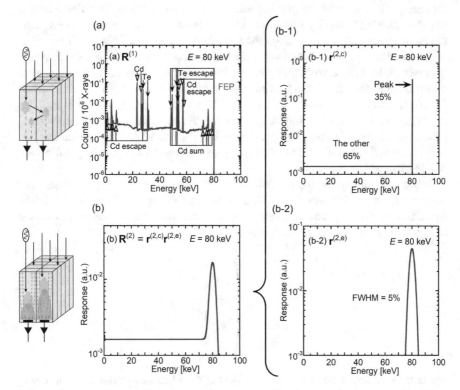

Fig. 5 Typical results of detector responses for 80 keV monochromatic X-ray. (**a**) $\mathbf{R}^{(1)}$ including phenomenon related to radiation interaction; the red and blue lines show full energy peak and scattered X-rays, respectively. (**b**) $\mathbf{R}^{(2)}$ including phenomenon related to the electric charge collection processes which are composed of (**b-1**) charge-sharing effect $\mathbf{r}^{(2,\,c)}$ and (**b-2**) energy resolution $\mathbf{r}^{(2,\,e)}$

represents the full energy peak (FEP) which appears when all of the incident X-ray energy is completely absorbed by the pixel of interest. On the other hand, the blue area relates to partially absorbed events, and some peaks are observed at 23–32 and 48–57 keV. It is important to understand these phenomena; when the X-rays are incident to the pixel, the photoelectric effect mainly occurs and then characteristic X-rays of Cd (23–27 keV) and Te (27–32 keV) are generated. When focusing on the pixel of interest, there is a possibility of the characteristic X-rays escaping to adjacent pixels, which results in the existence of escape peaks (EPs) at 48–57 keV. On the other hand, when focusing on the pixel adjacent to the pixel of interest, the characteristic X-rays generating at pixel of interest are incident, which results in the existence of peaks at 23–32 keV. It should be noted that the characteristic X-ray peaks are constant regardless of the energy of the incident X-rays, whereas the EPs vary according to the energy of the incident X-rays. Next, the response function $\mathbf{R}^{(2)}$ is reproduced taking into consideration charge sharing effect $\mathbf{r}^{(2,\,c)}$ and energy resolution $\mathbf{r}^{(2,\,e)}$. During charge collection process in a detector, electrons undergo a diffusion process and drift due to an electric field. Consequently, not all of the charges are collected by the pixel of interest and some charges spread into several adjacent pixels caused by diffusions and collisions of produced electrons. This phenomenon is called the "charge sharing effect." One vector element in $\mathbf{r}^{(2,\,c)}$ is exemplified in Fig. 5b-1. It consists of the peak component and the other part which is presented by a flat distribution. For our ERPCD, the respective proportions were optimized and the values were determined to be 35% and 65% so as to reproduce the X-ray spectrum measured with the ERPCD. In addition, as shown in Fig. 5b-2, energy resolution $\mathbf{r}^{(2,\,e)}$ was determined as 5% at 80 keV so as to reproduce characteristic X-ray peaks. Using $\mathbf{r}^{(2,\,c)}$ and $\mathbf{r}^{(2,\,e)}$, the 80 keV vector element in $\mathbf{R}^{(2)}$ is exemplified in Fig. 5b in which the peaks around 80 keV and flat distribution can be observed.

Next, we will explain the procedure to reproduce an X-ray spectrum using the response functions $\mathbf{R}^{(1)}$ and $\mathbf{R}^{(2)}$. To easily understand the procedure, we reproduce two different X-ray spectra obtained under the ideal and actual situations as shown in Fig. 4. First, we set the X-ray spectrum,

$$\mathbf{\Phi} = \begin{pmatrix} \Phi_1 \\ \Phi_2 \\ \vdots \\ \Phi_j \\ \vdots \end{pmatrix}, \tag{2}$$

where the element of this vector is expressed as Φ_j. Figure 6a shows $\mathbf{\Phi}$ when setting the tube voltage at 50 kV.

In the ideal case, the response function can be expressed as the identity matrix:

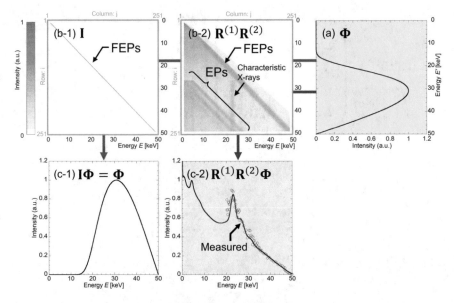

Fig. 6 The procedure for reproducing an X-ray spectrum taking into consideration detector response. (**a**) Incident X-ray spectrum $\mathbf{\Phi}$. (**b-1**) and (**b-2**) show two-dimensional color maps of detector responses $\mathbf{I}$ and $\mathbf{R}^{(1)}\mathbf{R}^{(2)}$, respectively. (**c-1**) and (**c-2**) show reproduced X-ray spectra $\mathbf{I\Phi}$ $= \mathbf{\Phi}$ and $\mathbf{R}^{(1)}\mathbf{R}^{(2)}\mathbf{\Phi}$, respectively. By performing the mathematical calculations, we can reproduce X-ray spectra measured with an ERPCD

$$
\mathbf{I} = \begin{pmatrix} 1 & 0 & \cdots\cdots & 0 \\ 0 & 1 & \vdots & \vdots & \vdots \\ \vdots & \vdots & \ddots & \vdots & \vdots \\ \vdots & \vdots & \vdots & 1 & 0 \\ 0 & \cdots\cdots & & 0 & 1 \end{pmatrix}.
\tag{3}
$$

Figure 6b-1 shows a two-dimensional color map presented by $\mathbf{I}$. It can be seen that unit values corresponding to FEPs appear in diagonal line.

In actual circumstances, the response function can be calculated by the matrix operation for $\mathbf{R}^{(1)}$ and $\mathbf{R}^{(2)}$,

$$\mathbf{R}^{(1)}\mathbf{R}^{(2)} = \begin{pmatrix} R_{1,1}^{(1)} & R_{1,2}^{(1)} & \cdots & \cdots\cdots \\ R_{2,1}^{(1)} & \ddots & \vdots & \vdots & \vdots \\ \vdots & \vdots & R_{i,k}^{(1)} & \vdots & \vdots \\ \vdots & \vdots & \vdots & \ddots & \vdots \\ \cdots & \cdots & \cdots\cdots \end{pmatrix} \begin{pmatrix} R_{1,1}^{(2)} & R_{1,2}^{(2)} & \cdots & \cdots\cdots \\ R_{2,1}^{(2)} & \ddots & \vdots & \vdots & \vdots \\ \vdots & \vdots & R_{k,j}^{(2)} & \vdots & \vdots \\ \vdots & \vdots & \vdots & \ddots & \vdots \\ \cdots & \cdots & \cdots\cdots \end{pmatrix}$$

$$= \begin{pmatrix} R_{1,1}^{(1,2)} & R_{1,2}^{(1,2)} & \cdots & \cdots\cdots \\ R_{2,1}^{(1,2)} & \ddots & \vdots & \vdots & \vdots \\ \vdots & \vdots & R_{i,j}^{(1,2)} & \vdots & \vdots \\ \vdots & \vdots & \vdots & \ddots & \vdots \\ \cdots & \cdots & \cdots & \cdots\cdots \end{pmatrix},$$

(4)

where $R_{i,j}^{(1,2)}$ is described as

$$R_{i,j}^{(1,2)} = \sum_k R_{i,k}^{(1)} R_{k,j}^{(2)}.$$

$\mathbf{R}^{(1)}$ consists of the elements of $\mathbf{R}_{E'}^{(1)}$ corresponding to various incident X-ray energies E'. Then, $\mathbf{R}_{E'}^{(1)}$ is expressed as $\left\{ R_{i,1}^{(1)}, R_{i,2}^{(1)}, \ldots, R_{i,k}^{(1)}, \ldots \right\}$ where the corresponding response energy is the i-th element. In a similar way, $\mathbf{R}^{(2)}$ has elements of $\mathbf{R}_{E'}^{(2)}$. Then, the element $\mathbf{R}^{(1)}\mathbf{R}^{(2)}$ in the i-th row and j-th column is expressed as $R_{i,j}^{(1,2)}$. Figure 6b-2 shows a two-dimensional color map of $\mathbf{R}^{(1)}\mathbf{R}^{(2)}$. The response corresponding to FEPs is shown with the diagonal line and the sharpness is affected by the energy resolution. The characteristic X-rays are observed at around 23 keV. EPs are represented by diagonal lines where the highest value is around 27 keV. The charge-sharing effect causes an increase in intensities in the low-energy region.

Finally, we will reproduce X-ray spectra using response functions $\mathbf{I}$ and $\mathbf{R}^{(1)}\mathbf{R}^{(2)}$. In ideal case, X-ray spectrum can be obtained by folding Φ with $\mathbf{I}$,

$$\mathbf{I}\Phi = \begin{pmatrix} 1 & 0 & \cdots\cdots & 0 \\ 0 & 1 & \vdots & \vdots & \vdots \\ \vdots & \vdots & \ddots & \vdots & \vdots \\ \vdots & \vdots & \vdots & 1 & 0 \\ 0 & \cdots\cdots & 0 & 1 \end{pmatrix} \begin{pmatrix} \Phi_1 \\ \Phi_2 \\ \vdots \\ \Phi_j \\ \vdots \end{pmatrix} = \begin{pmatrix} \Phi_1 \\ \Phi_2 \\ \vdots \\ \Phi_j \\ \vdots \end{pmatrix}.$$

(5)

Figure 6c-1 shows $\mathbf{I}\Phi$. This is the same as incident X-ray spectrum Φ.

In the actual case, we can reproduce the X-ray spectrum by replacing $\mathbf{I}$ in Eq. (5) by $\mathbf{R}^{(1)}\mathbf{R}^{(2)}$:

$$\mathbf{R}^{(1)}\mathbf{R}^{(2)}\mathbf{\Phi} = \begin{pmatrix} R_{1,1}^{(1,2)} & R_{1,2}^{(1,2)} & \cdots & \cdots & \cdots \\ R_{2,1}^{(1,2)} & \ddots & \vdots & \vdots & \vdots \\ \vdots & \vdots & R_{i,j}^{(1,2)} & \vdots & \vdots \\ \vdots & \vdots & \vdots & \ddots & \vdots \\ \cdots & \cdots & \cdots & \cdots & \cdots \end{pmatrix} \begin{pmatrix} \Phi_1 \\ \Phi_2 \\ \vdots \\ \Phi_j \\ \vdots \end{pmatrix} = \begin{pmatrix} \sum_j R_{1,j}^{(1,2)}\Phi_j \\ \sum_j R_{2,j}^{(1,2)}\Phi_j \\ \vdots \\ \sum_j R_{i,j}^{(1,2)}\Phi_j \\ \vdots \end{pmatrix},$$

(6)

where each element is described as

$$\sum_j R_{i,j}^{(1,2)}\Phi_j = \sum_j \left(\sum_k R_{i,k}^{(1)} R_{k,j}^{(2)} \right) \Phi_j = \sum_{E'} \left(\sum_k R_{i,k}^{(1)} R_{k,E'}^{(2)} \right) \Phi\left(E'\right),$$

where j is the j-th element of $\mathbf{\Phi}$, for the sake of clarification, j is expressed as E'. Figure 6c-2 shows the reproduced X-ray spectrum $\mathbf{R}^{(1)}\mathbf{R}^{(2)}\mathbf{\Phi}$ plotted with the X-ray spectrum measured using our proto-type ERPCD [13, 15, 38]. $\mathbf{R}^{(1)}\mathbf{R}^{(2)}\mathbf{\Phi}$ is in good agreement with the experimental data. We can see that the $\mathbf{R}^{(1)}\mathbf{R}^{(2)}\mathbf{\Phi}$ has two major features: relatively large intensities in the low-energy region and the presence of the characteristic X-ray peaks of Cd and Te. The $\mathbf{R}^{(1)}\mathbf{R}^{(2)}\mathbf{\Phi}$ is much different than that of $\mathbf{\Phi}$.

Here, in order to simplify the mathematical notation, $\mathbf{\Phi}$ is redefined as follows: $\mathbf{\Phi} = \{\Phi(0 \text{ keV}), \Phi(0.2 \text{ keV}), \ldots, \Phi(E_k), \ldots\}$ where E_k is X-ray energy at intervals of 0.2 keV (subscript of "k" means k-th vector element) for matrix formation and $\Phi(E)$ for continuous function. We also redefine $\mathbf{R}^{(1)}\mathbf{R}^{(2)}\mathbf{\Phi}$ as matrix formation: $\mathbf{R}^{(1)}\mathbf{R}^{(2)}\mathbf{\Phi} = \{R^{(1)}R^{(2)}\Phi(0 \text{ keV}), R^{(1)}R^{(2)}\Phi(0.2 \text{ keV}), \ldots, R^{(1)}R^{(2)}\Phi(E_k), \ldots\}$. These formation roles are applied to all equations in the following description.

3.3 Procedure to Correct for the Beam Hardening Effect and Detector Response

To accomplish highly accurate material identification, the biggest challenges that need solving are the issues of the beam hardening effect and detector response. In this section, we will introduce the correction procedure for the beam hardening effect and detector response. Figure 7 shows a schematic drawing that explains the concept of the correction. The correction is to convert $(\mu t)_{meas}$ which is calculated from polychromatic X-rays detected by the ERPCD into $(\mu t)_{cor}$ calculated from monochromatic X-rays. To easily understand the correction of both effects, we focus on (a) monochromatic X-rays in $\mathbf{\Phi}$, (b) polychromatic X-rays in $\mathbf{\Phi}$, and (c) polychromatic X-rays in $\mathbf{R}^{(1)}\mathbf{R}^{(2)}\mathbf{\Phi}$. Inset (b) includes the beam hardening effect and (c) includes both beam hardening effect and detector response. As shown, common for conditions (a) through (c), we set the lower and upper energies of each energy bin as E_1 and E_2, respectively, and $\overline{E}$ is calculated. Assuming that

Correction method for the beam hardening effect and detector response

$$(\mu t)_{\text{meas}} \longrightarrow (\mu t)_{\text{cor}}$$

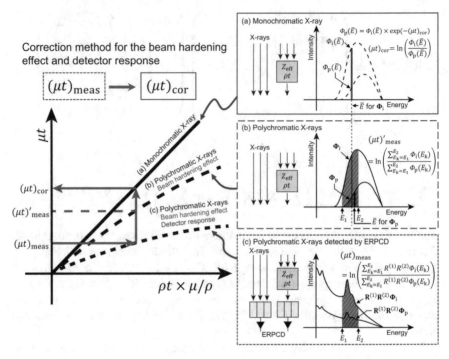

Fig. 7 Concept of the procedure to correct both the beam hardening effect and detector response. The correction is to convert $(\mu t)_{\text{meas}}$ calculated from polychromatic X-rays measured with an ERPCD to $(\mu t)_{\text{cor}}$ which is related to monochromatic X-rays. The former case consists of energy information disruption caused by the beam hardening effect and detector response, and the latter case is the retention of original energy information reflecting X-ray attenuation of the object. To clearly show the correction for both effects, we present each step of the correction: (**a**) $(\mu t)_{\text{cor}}$ calculated from monochromatic X-rays, (**b**) $(\mu t)'_{\text{meas}}$ calculated from polychromatic X-rays, and (**c**) $(\mu t)_{\text{meas}}$ calculated from polychromatic X-rays measured with the ERPCD

an object with certain Z_{eff} and ρt values is measured, we simulate the X-ray spectra and calculate corresponding μt values related to conditions (a) through (c). Then, we derive unique relationships between $\rho t \times \mu/\rho$ and μt which were proposed in a previous study [12, 14–16, 19].

Here, we explain the relationship of (a) monochromatic X-ray energy $\overline{E}$. The $(\mu t)_{\text{cor}}$ is calculated from the equation, $\ln\left(\dfrac{\Phi_i(\overline{E})}{\Phi_p(\overline{E})}\right)$. Because the $\overline{E}$ is always constant even when ρt and/or Z_{eff} are varied, $(\mu t)_{\text{cor}}$ values can be plotted as a straight line $Y = X$; X and Y axes are $\rho t \times \mu/\rho$ and μt, respectively. Next, for (b) polychromatic X-rays, because the $\overline{E}$ of Φ_p, which is calculated from the equation $\dfrac{\sum_{E_k=E_1}^{E_2} \Phi_p(E_k) E_k}{\sum_{E_k=E_1}^{E_2} \Phi_p(E_k)}$, is higher due to the beam hardening effect, the

$(\mu t)'_{\text{meas}}$ calculated from the equation $\ln\left(\dfrac{\sum_{E_k=E_1}^{E_2} \Phi_i(E_k)}{\sum_{E_k=E_1}^{E_2} \Phi_p(E_k)}\right)$ differs from the $(\mu t)_{\text{cor}}$; the $(\mu t)'_{\text{meas}}$ shows a curve which goes away from the $Y = X$ line. Finally, when using (c) polychromatic X-rays detected using an ERPCD, $(\mu t)_{\text{meas}}$ calculated from the equation $\ln\left(\dfrac{\sum_{E_k=E_1}^{E_2} R^{(1)} R^{(2)} \Phi_i(E_k)}{\sum_{E_k=E_1}^{E_2} R^{(1)} R^{(2)} \Phi_p(E_k)}\right)$ is much different from the $Y = X$ line due to the beam hardening effect and detector response. The curve was calculated at intervals of 0.1 g/cm^2 for virtual objects having $Z_{\text{eff}} = 5.0$–15.0 with $\rho t = 0$–150 g/cm^2.

Based on the relationship between (a) and (c), the $(\mu t)_{\text{meas}}$ measured with ERPCD can be converted into $(\mu t)_{\text{cor}}$ as shown by the direction of red arrows in the left panel. This correction means that both effects can be corrected simultaneously and polychromatic X-rays can be treated as monochromatic X-rays. Although we can apply the correction accurately under the assumption in which the Z_{eff} of an object is known, it is a rare case in advance. To make the best use of the correction curve, we propose to determine Z_{eff} while estimating the amount of correction in later sections.

4 Material Identification Method Leading to Innovations in Imaging Technology

In this section, we would like to introduce our material identification method. The biggest advantage of our method is that the beam hardening effect and detector response can be corrected appropriately by taking into consideration the Z_{eff} of an object. This correction realizes the ideal analysis in which polychromatic X-rays measured with the ERPCD can be treated as those equivalent to monochromatic X-rays. As an application example showing the effectiveness of this analysis, we succeeded not only in identifying the Z_{eff} of an object but also in extracting ρt related to soft-tissue and bone. In the following section, we introduce the proposed method to derive Z_{eff} and ρt values of soft tissue and bone. We then show the feasibility of our method visually using images produced with prototype ERPCD.

4.1 Novel Method to Derive Effective Atomic Number Image and Extract Mass Thickness Images Related to Soft Tissue and Bone

We will introduce a method to derive the Z_{eff} and extract the ρt values related to soft tissue and bone. Figure 8 shows a diagram of the method. Assuming that a bilayer structure consisting of "a" and "b" is measured using an ERPCD having several energy bins, we perform the experiments without and with the presence of objects, as shown in the upper center panel. In the current system, we adopted to use two energy bins. Steps Fig. 8a–f are applied to the information of these energy bins to derive Z_{eff} value and each ρt value. First, we prepare count values which are measured in the experiments without and with objects as shown in Fig. 8a. The count values without and with objects correspond to mathematical expressions $\sum_{E_k=E_1}^{E_2} R^{(1)} R^{(2)} \Phi_i (E_k)$ and $\sum_{E_k=E_1}^{E_2} R^{(1)} R^{(2)} \Phi_p (E_k)$, respectively. Next, using these values, attenuation factor $(\mu t)_{meas}$ is calculated as

$$(\mu t)_{meas} = \ln \left(\frac{\sum_{E_k=E_1}^{E_2} R^{(1)} R^{(2)} \Phi_i (E_k)}{\sum_{E_k=E_1}^{E_2} R^{(1)} R^{(2)} \Phi_p (E_k)} \right). \tag{7}$$

This calculation is applied to low- and high-energy bins, and $(\mu_{low} t)_{meas}$ and $(\mu_{high} t)_{meas}$ are derived as shown in Fig. 8b. Because $(\mu t)_{meas}$ is affected by the beam hardening effect and detector response, corrections related to various Z_{tent} values which are the values tentatively assigned as the Z_{eff} of an object are performed in each energy bin; namely, $(\mu t)_{meas}$ is converted into $(\mu t)_{cor}$ values related to Z_{tent} values 5, 6, 7, $\ldots$, 15, as shown in Fig. 8c. Then, to extract μ using $(\mu_{low} t)_{cor}$ and $(\mu_{high} t)_{cor}$, the following calculation is performed:

$$\mu^\dagger \equiv \frac{(\mu_{high} t)_{cor}}{\sqrt{(\mu_{low} t)_{cor}^2 + (\mu_{high} t)_{cor}^2}} = \frac{(\mu_{high})_{cor}}{\sqrt{(\mu_{low})_{cor}^2 + (\mu_{high})_{cor}^2}}, \tag{8}$$

where $\mu^\dagger$ is a normalized attenuation factor [8, 10–12, 14–16, 18, 19], and we obtain $\mu^\dagger$ values related to Z_{tent} values. In order to derive Z_{eff} from $\mu^\dagger$, a theoretical relationship between $\mu^\dagger$ and Z_{eff} is created using database μ/ρ [24]. Using this reference curve, experimentally derived $\mu^\dagger$ values can be converted into the corresponding Z_{eff} values as shown in Fig. 8d. The derived Z_{eff} values are plotted as a function of Z_{tent}, and Z_{eff} of an object can be determined by searching for the intersectional point between the derived Z_{eff} curve and $Y = X$ line. This algorithm to determine the Z_{eff} of an object takes advantage of the following trend. When Z_{tent} is the same as Z_{eff} of an object, the correction can perform appropriately, and the derived Z_{eff} becomes Z_{eff} of an object. On the other hand, when Z_{tent} is different from Z_{eff} of an object, the derived Z_{eff} is also different from Z_{eff} of an object due to improper correction. This means that the intersectional point of the Z_{tent} versus Z_{eff} plot indicates the Z_{eff} of an object, and we can determine Z_{eff} of the object using this algorithm.

Furthermore, taking advantage of the fact that the Z_{eff} of an object can be determined and simultaneously appropriate correction can be applied, we can perform further analysis and extract other properties of an object. We focus on the use of $(\mu t)_{cor}$ with the correction when obtaining the Z_{eff} of an object. Fortunately, when going back to Fig. 8c, a database of $(\mu t)_{cor}$ values related to Z_{tent} values has been obtained. Now plotting $(\mu t)_{cor}$ as a function of Z_{tent}, the $(\mu t)_{cor}$ related to Z_{eff}

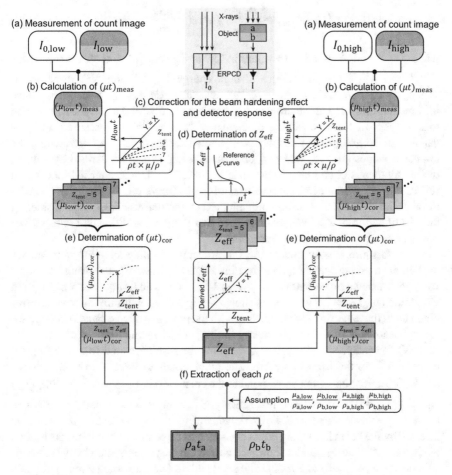

Fig. 8 Diagram of the proposed algorithm for determining the Z_{eff} and ρt values. Assuming that a bilayer structure consisting of elements "a" and "b" is measured using an ERPCD, we perform the experiments without and with the presence of the object. The six steps (a) through (f) are applied to measurement data from two energy bins to derive the Z_{eff} value and the ρt values corresponding to soft tissue and bone

can be determined as shown in Fig. 8e. Because the $(\mu t)_{cor}$ related to Z_{eff} shows that the beam hardening effect and detector response can be corrected completely, ideal analysis based on monochromatic X-ray energy can be carried out. As one valuable application using $(\mu t)_{cor}$, we can extract ρt values of each element by solving simultaneous equations:

$$(\mu_{low}t)_{cor} = \frac{\mu_{a,low}}{\rho_a}\rho_a t_a + \frac{\mu_{b,low}}{\rho_b}\rho_b t_b, \tag{9}$$

$$\left(\mu_{\text{high}}t\right)_{\text{cor}} = \frac{\mu_{a,\text{high}}}{\rho_a}\rho_a t_a + \frac{\mu_{b,\text{high}}}{\rho_b}\rho_b t_b, \tag{10}$$

where $\frac{\mu_{a,\text{low}}}{\rho_a}$ and $\frac{\mu_{b,\text{low}}}{\rho_b}$ are μ/ρ values at low energy for "a" and "b", respectively, $\frac{\mu_{a,\text{high}}}{\rho_a}$ and $\frac{\mu_{b,\text{high}}}{\rho_b}$ are μ/ρ values at high energy for "a" and "b", respectively, and $\rho_a t_a$ and $\rho_b t_b$ are ρt values of "a" and "b", respectively. Because we have information related to monochromatic X-ray energy ($\overline{E}$ of Φ_i) under the assumption that bilayer structure consisting of elements "a" and "b" is measured, the μ/ρ values can be assigned by using a well-known database [24]. Consequently, the unknown values are $\rho_a t_a$ and $\rho_b t_b$ and, therefore, they can be derived by solving the simultaneous equations as shown in Fig. 8f. This is one analysis that makes full use of our method in which monochromatic X-rays can be virtually obtained through the correction of beam hardening effect and detector response considering the identification of Z_{eff}. We are convinced that there are additional possibilities of discovering novel values using our method.

Next, to more clearly understand the procedure to determine Z_{eff} and ρt, we explain each process using typical results. Figure 9 shows typical results obtained when measuring a bilayer structure ($Z_{\text{eff}} = 10.5$) of acrylic ($\rho_{\text{ac}} t_{\text{ac}} = 2.5$ g/cm^2) and aluminum ($\rho_{\text{Al}} t_{\text{Al}} = 2.5$ g/cm^2). As shown in Fig. 9a, we measure the count images obtained in the experiments without and with the presence of an object. Although we use count images from the low-energy bin (32–40 keV) and high-energy bin (40–50 keV) for analysis, for simplicity, we show the results related to a certain pixel of images from the low-energy bin. The pixel values of interest in count images are 50,270 and 4506 which correspond to the calculations $\sum_{E_k=E_1}^{E_2} R^{(1)} R^{(2)} \Phi_i(E_k)$ and $\sum_{E_k=E_1}^{E_2} R^{(1)} R^{(2)} \Phi_p(E_k)$, respectively. Substituting these values into Eq. (7), $(\mu_{\text{low}}t)_{\text{meas}}$ is calculated as 2.41 as shown in Fig. 9b. Because the $(\mu_{\text{low}}t)_{\text{meas}}$ is affected by the beam hardening effect and detector response, correction is performed as shown in Fig. 9c. Figure 9c shows the correction curves presented by the relationship between $\rho t \times \mu/\rho$ and $\mu_{\text{low}}t$. The solid and broken lines show monochromatic X-rays and polychromatic X-rays, respectively. In this graph, monochromatic X-rays are plotted on the $Y = X$ line regardless of Z_{tent}, and the polychromatic X-ray trend depends on Z_{tent}. The difference between monochromatic X-rays and polychromatic X-rays indicates the amount of correction; it is seen that at higher Z_{tent}, the amount of correction is larger. By assuming that the Z_{tent} values are between 5 and 15, the $(\mu_{\text{low}}t)_{\text{meas}}$ can be converted into $(\mu_{\text{low}}t)_{\text{cor}}$ values which correspond to Z_{tent} values. In this demonstration of a typical case shown in Fig. 9c, when we set $Z_{\text{tent}} = 15$, $(\mu_{\text{low}}t)_{\text{meas}} = 2.41$ is converted into $(\mu_{\text{low}}t)_{\text{cor}} = 2.66$, as represented by the red arrows. These corrections are also applied to high-energy bin to obtain $(\mu_{\text{high}}t)_{\text{cor}}$ values. In order to determine Z_{eff}, $\mu^{\dagger}$ values are calculated by substituting $(\mu_{\text{low}}t)_{\text{cor}}$ and $(\mu_{\text{high}}t)_{\text{cor}}$ values to Eq. (8). The upper figure in Fig. 9d shows the relationship between $\mu^{\dagger}$ and Z_{eff}. The broken line shows the reference curve determined using a well-known database [24]. As shown by the arrows, $\mu^{\dagger} = 0.57$ corresponding to $Z_{\text{tent}} = 15$ can be converted to $Z_{\text{eff}} = 11.5$. To determine Z_{eff} of an object, the derived Z_{eff} values are plotted as a function of Z_{tent}

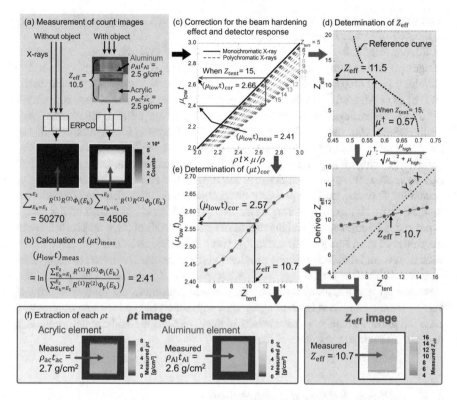

Fig. 9 Typical results to explain each process to derive Z_{eff} and ρt images. Using the data measured for a bilayer structure consisting of acrylic and aluminum, the Z_{eff} and ρt images can be derived by performing the six steps (a) through (f)

as shown in the lower figure in Fig. 9d. By searching for the intersectional point between the derived Z_{eff} curve and $Y = X$ line, the Z_{eff} of an object is determined to be 10.7. The determined Z_{eff} is in good agreement with $Z_{\text{eff}} = 10.5$ which is predetermined as a reference value using a well-known formula [30].

Furthermore, we want to determine ρt by taking advantage of the Z_{eff}. Because $(\mu_{\text{low}}t)_{\text{cor}}$ values related to Z_{tent} values are already obtained in the correction process as shown in Fig. 9c, the $(\mu_{\text{low}}t)_{\text{cor}}$ is plotted as a function of Z_{tent} as shown in Fig. 9e. Then, the $(\mu_{\text{low}}t)_{\text{cor}}$ related to an $Z_{\text{eff}} = 10.7$ can be determined as 2.57. Because the determined $(\mu_{\text{low}}t)_{\text{cor}}$ has completely corrected the beam hardening effect and detector response, we can perform the following analysis which can be carried out under the limited conditions of being able to use monochromatic X-rays. Substituting $(\mu_{\text{low}}t)_{\text{cor}}$ and $(\mu_{\text{low}}t)_{\text{cor}}$ values into Eqs. (9) and (10) under the assumption that the object consists of acrylic and aluminum, namely, μ/ρ values are known, the $\rho_{\text{ac}}t_{\text{ac}}$ and $\rho_{\text{Al}}t_{\text{Al}}$ can be determined to be 2.7 g/cm^2 and 2.6 g/cm^2, respectively, as shown in Fig. 9f. The determined ρt values are in good agreement with reference ρt values of 2.5 g/cm^2. These procedures are applied pixel by pixel to create Z_{eff} and ρt images.

To confirm that our method can determine Z_{eff} and ρt values, we present images produced using our prototype ERPCD. Figure 10 shows images of acrylic, aluminum, and bilayer structures of acrylic and aluminum having (a) $\rho t = 1.0$ g/cm^2 and (b) $\rho t = 5.0$ g/cm^2. As shown in the photograph, acrylic, aluminum, and bilayer structures are arranged in order from left to right. In our system, the conventional X-ray image, Z_{eff} image, and ρt image can be obtained. In conventional X-ray images which are equivalent to the image generated by EID, the image density varies depending on Z_{eff} and ρt. When the Z_{eff} is higher with the same ρt, the image density is smaller. Also, when the ρt is larger with the same Z_{eff}, the image density is smaller. It means that the property of an object cannot be identified because the image density changes depending on the two parameters, Z_{eff} and ρt. The Z_{eff} image is presented in color scale, and the region of interest (ROI) of 50×50 pixels is set in the center of each sample. The mean Z_{eff} value measured in ROI is presented below each image, and theoretical Z_{eff} is provided in parentheses. The measured Z_{eff} value agrees with the theoretical Z_{eff} with an accuracy of $Z_{eff} \pm 0.4$. In the ρt images, elements for acrylic, aluminum, and acrylic and aluminum are presented. The ρt value for each sample is in good agreement with theoretical ρt with an accuracy of $\rho t \pm 0.5$ g/cm^2. Therefore, these results indicate that our method can determine Z_{eff} and extract each ρt regardless of the sample.

4.2 Spillover Effect on Applications

In this section, we will discuss the usefulness of our research by visually presenting images measured with our proto-type ERPCD. As generally known, biological tissue such as the human body, and food are often composed of soft tissue and bone, and images corresponding to their separation will help achieve various diagnoses through examination. In the radiography of the lung area, images with bones removed are in great demand to confirm the presence of lesions [39–41]. In the mammography examination, identification of calcification buried in soft tissue is also necessary [42]. Furthermore, our system can accurately measure the ρt of bone from a bone image. The ρt of bone is synonymous with bone mineral density (BMD) which is used as a diagnostic index for osteoporosis, which indicates that our system has the potential to diagnose osteoporosis without using specific equipment (dual-energy X-ray absorptiometry) [43–45]. These are just a few examples of how images can be useful. We think there are various fields in which Z_{eff} and ρt images can be used effectively.

Figure 11 shows the results of an experiment using a panoramic dental radiograph [46, 47]. Phantom consisting of actual teeth and jawbone and BMD calibrators were analyzed using a prototype ERPCD system, as shown in the photograph. Jawbones are dried human bones. The image on the upper left is laid out with only the bones

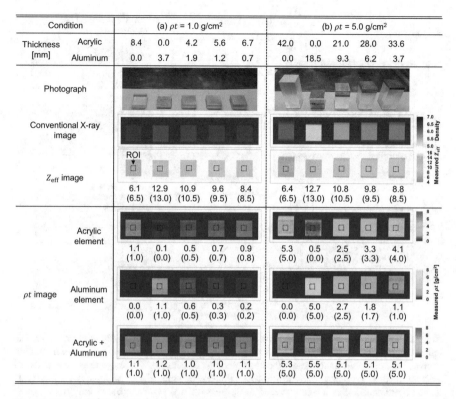

Condition		(a) ρt = 1.0 g/cm²					(b) ρt = 5.0 g/cm²				
Thickness [mm]	Acrylic	8.4	0.0	4.2	5.6	6.7	42.0	0.0	21.0	28.0	33.6
	Aluminum	0.0	3.7	1.9	1.2	0.7	0.0	18.5	9.3	6.2	3.7

Fig. 10 The results for verifying our novel method to derive Z_{eff} and ρt images using a proto-type ERPCD system. The upper row of images shows photographs of samples having (a) $\rho t = 1.0$ g/cm² and (b) $\rho t = 5.0$ g/cm². The acrylic, aluminum, and bilayer structures of acrylic and aluminum are arranged in order from left to right. The second row of images shows conventional X-ray images. The third row of images shows Z_{eff} images, and the ROI of 50 × 50 pixels is set in each image. The mean Z_{eff} value measured in ROI is presented below each image, and the theoretical Z_{eff} value is also presented in parentheses. The last three rows of images show ρt images for the acrylic element, the aluminum element, and both the acrylic and aluminum elements. The mean ρt value and theoretical value are presented below. In the Z_{eff} and ρt images, the measured values are in good agreement with theoretical values

exposed. That on the right is placed on the acrylic plate (1 cm); this assumes actual dental radiography. In the actual situation, an X-ray photograph is taken with X-rays that pass through both the skin and teeth and/or bones, and it is difficult to perform a precise quantitative analysis of the bones because the attenuation information related to skin (soft tissue) is superimposed on the teeth and/or bones. We believe that the soft tissue and bone separation techniques introduced in this chapter will be useful. One of the analyses related to the jawbone is the calculation of BMD (ρt). In order to show that BMD can be calculated through dental imaging, an experiment was performed with BMD calibrators [48–50]. The bar phantom on the left is a clinically used calibrator (UHA-type BMD phantom, Kyoto Kagaku Co., Ltd.,

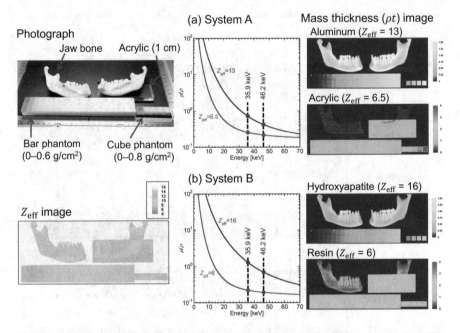

Fig. 11 Analysis results of teeth and jaw bones to demonstrate the feasibility of our method. The left photograph shows the experimental arrangement, in which phantoms and a BMD calibrator are shown. The below-left image shows the corresponding Z_{eff} image. The image data were analyzed with our method to extract bone and soft tissue images, in which Z_{eff} of two material components should be assumed. The image was then analyzed with two systems: (**a**) system "A" in which $Z_{eff} = 13$ (aluminum) and 6.5 (acrylic) were assumed and (**b**) system "B" in which $Z_{eff} = 16$ (hydroxyapatite) and 6 (resin) were assumed

Japan) that can calibrate BMD from 0 to 0.6 g/cm^2. This phantom is usually used for general radiography of the hand [51, 52]. The cube phantom on the right was made by mixing hydroxyapatite with epoxy-acrylic resin, and the range of BMD was 0–0.8 g/cm^2. In the Z_{eff} image shown on the lower left, it can be seen that the Z_{eff} values of the teeth and jawbones vary depending on the presence of the acrylic plate and that the Z_{eff} of the BMD phantom changes gradually. Since the Z_{eff} image reflects all material information in the X-ray penetration path of the object, it does not provide much important information for this demonstration. However, when extracting the bone and soft-tissue components, information concerning Z_{eff} is required as shown in Fig. 8. As mentioned above, in order to extract bone and soft tissue, it is necessary to assume the mass energy absorption coefficient (and/or related Z_{eff} values) of bone and soft tissue. Here, we performed verification experiments under the following two systems. Figure 11a is system "A" in which $Z_{eff} = 13$ (aluminum) and 6.5 (acrylic) were applied. Figure 11b is system "B" in which $Z_{eff} = 16$ (hydroxyapatite) and 6 (resin) were used. It can be seen that the mass energy absorption coefficients are different for systems A and B. Two energy values of 35.9 and 46.2 keV are effective energies of low- and high-energy bins. The mass thickness images of high

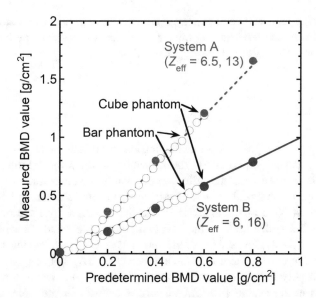

Fig. 12 Relationship between predetermined BMD values and measured ones. Although systems A and B (see Fig. 11) show good linear correlations, good agreement between them can be obtained only when system B is applied. Using system B, our ERPCD can determine BMD value without a BMD calibrator

and low Z_{eff} elements were different for systems A and B, but it is difficult to identify the differences quantitatively when looking at the color image. On the other hand, it was clearly seen that the jawbone area does not contain the acrylic component that mimics the soft tissue. This analysis cannot be performed in the current dental radiography system using a general EID, and by using ERPCD with our algorithm, high-quality diagnostic information can be obtained.

Figure 12 shows the results of the quantification analysis of the mass thickness images of bone calculated in Fig. 11. The horizontal axis is the predetermined values of hydroxyapatite contained in the bar and cube phantoms, and the vertical axis is the measured BMD values calculated from the mass thickness image. Both systems A and B show good linearity for the X and Y axes, which means that BMD can be measured using the ERPCD. System B indicated an $X = Y$ correlation, which means that absolute BMD measurement can be performed without using a BMD calibrator. The phantoms used in this experiment are composed of a mixture of hydroxyapatite and resin and these two components could be completely separated by system B. It should be noted that part of the jawbone was depicted in the resin image of system B ($Z_{eff} = 6$), even though the jawbone was not present in the acrylic image of system A ($Z_{eff} = 6.5$). We cannot deny the possibility that the jawbone sample contains a small amount of a component related to resin; therefore, we cannot conclude that the resin image of system B was incorrect. In this experiment, the atomic number of the BMD phantom region that does not contain hydroxyapatite was 6, so extraction can be performed using $Z_{eff} = 6$ and the theoretical value of hydroxyapatite ($Z_{eff} = 16$)

[53]. In the future, it will be necessary to conduct similar trials using actual clinical examinations. From these trials, optimization of the system can be performed using the actual measured values of the Z_{eff} of soft tissue.

5   Summary

In this chapter, an algorithm for extracting the properties of an object using an ERPCD was described. Based on the principle in which the information of the object can be derived from the differences in the X-ray attenuation depending on X-ray energy, we analyzed polychromatic X-rays measured with an ERPCD. The most important issue of the analysis is that the measured information related to each energy bin is distorted by the beam hardening effect and detector response. We described these effects from a viewpoint based on physics and presented a methodology to correct these effects simultaneously. By applying this correction, we can treat polychromatic X-rays as those equivalent to monochromatic X-rays. This results in the accomplishment of correct analysis in principle. In order to demonstrate our method, we succeeded in the derivation of a Z_{eff} image and extraction of soft-tissue and bone images by applying our algorithm to each pixel of an image measured with a proto-type ERPCD.

It should be noted that the experiments in this chapter were performed under ideal conditions, and our conclusion may not necessarily apply to actual CT systems. This is because the phantoms were measured at a low dose rate, and the effect of pulse pile-up was negligibly small. Furthermore, since the object and the detector distance are sufficiently separated, the influence of scattered X-rays can be almost ignored. When considering the actual equipment that the examination is being performed under clinical condition in which pulse pile-up and/or contamination of scattered X-rays occur, these effects must be included in the estimation process of the detector's response. The algorithm we presented is very simple in which information related to only two energy bins was used. If scientists want to apply our algorithm to other systems in which three or more energy bins are used, we recommend to modify the measured data; by summing the information of different energy bins, a two-energy-bin system can be easily achieved using a software approach.

We hope that the content described in this chapter opens up possibilities for the use of photon-counting techniques in medical and industrial fields.

Acknowledgments The description in this chapter was partially supported by collaborative research between Kanazawa University and JOB CORPORATION (https://www.job-image.com/), Japan. The authors wish to express gratitude to members of JOB CORPORATION, Dr. Shuichiro Yamamoto, Mr. Masahiro Okada, Mr. Fumio Tsuchiya, Mr. Daisuke Hashimoto, Mr. Yasuhiro Kuramoto, and Mr. Masashi Yamasaki for their valuable contributions. We wish to thank Mr. Takumi Asakawa, GE Healthcare Japan for his important research results when he belonged to graduate school in Kanazawa University, Japan. We would like to thank Dr. Yuki Kanazawa, Tokushima University, Japan, for discussing the feasibility of our research from a clinical point of view. We would also like to thank Dr. Yoshie Kodera and Dr. Shuji Koyama, Nagoya University, Japan, for discussing the clinical application of a photon-counting detector.

References

1. Pisano ED, et al. Image processing algorithms for digital mammography: a pictorial essay. Radiographics. 2000;20(5):1479–91. https://doi.org/10.1148/radiographics.20.5.g00se311479.
2. Prokop M, et al. Principles of image processing in digital chest radiography. J Thorac Imaging. 2003;18(3):148–64.
3. Kermany DS, et al. Identifying medical diagnoses and treatable diseases by image-based deep learning. Cell. 2018;172(5):1122–31. https://doi.org/10.1016/j.cell.2018.02.010.
4. Leng S, et al. Photon-counting detector CT: system design and clinical applications of an emerging technology. Radiographics. 2019;39(3):729–43. https://doi.org/10.1148/rg.2019180115.
5. Taguchi K, Iwanczyk JS. Vision 20/20: single photon counting x-ray detectors in medical imaging. Med Phys. 2013;40(10):100901. https://doi.org/10.1118/1.4820371.
6. Willemink MJ, et al. Photon-counting CT: technical principles and clinical prospects. Radiology. 2018;289(2):293–312. https://doi.org/10.1148/radiol.2018172656.
7. Asakawa T, et al. Importance of considering the response function of photon counting detectors with the goal of precise material identification. In: IEEE Nuclear Science Symposium and Medical Imaging Conference (NSS/MIC). IEEE; 2019. p. 1–7. https://doi.org/10.1109/NSS/MIC42101.2019.9059844.
8. Hayashi H, et al. A fundamental experiment for novel material identification method based on a photon counting technique: using conventional X-ray equipment. In: IEEE Nuclear Science Symposium and Medical Imaging Conference (NSS/MIC). IEEE; 2015. p. 1–4. https://doi.org/10.1109/NSSMIC.2015.7582027.
9. Hayashi H, et al. Response functions of multi-pixel-type CdTe detector: toward development of precise material identification on diagnostic X-ray images by means of photon counting. In: Medical imaging 2017: physics of medical imaging. SPIE; 2017. p. 1–18. https://doi.org/10.1117/12.2251185.
10. Kimoto N, et al. Precise material identification method based on a photon counting technique with correction of the beam hardening effect in X-ray spectra. Appl Radiat Isot. 2017;124:16–26. https://doi.org/10.1016/j.apradiso.2017.01.049.
11. Kimoto N, et al. Development of a novel method based on a photon counting technique with the aim of precise material identification in clinical X-ray diagnosis. In: Medical imaging 2017: physics of medical imaging. SPIE; 2017. p. 1–11. https://doi.org/10.1117/12.2253564.
12. Kimoto N, et al. Novel material identification method using three energy bins of a photon counting detector taking into consideration Z-dependent beam hardening effect correction with the aim of producing an X-ray image with information of effective atomic number. In: IEEE Nuclear Science Symposium and Medical Imaging Conference (NSS/MIC). IEEE; 2017. p. 1–4. https://doi.org/10.1109/NSSMIC.2017.8533059.
13. Kimoto N, et al. Reproduction of response functions of a multi-pixel-type energy-resolved photon counting detector while taking into consideration interaction of X-rays, charge sharing and energy resolution. In: IEEE Nuclear Science Symposium and Medical Imaging Conference (NSS/MIC). IEEE; 2018. p. 1–4. https://doi.org/10.1109/NSSMIC.2018.8824417.
14. Kimoto N, et al. Feasibility study of photon counting detector for producing effective atomic number image. In: IEEE Nuclear Science Symposium and Medical Imaging Conference (NSS/MIC). IEEE; 2019. p. 1–4. https://doi.org/10.1109/NSS/MIC42101.2019.9059919.
15. Kimoto N, et al. Effective atomic number image determination with an energy-resolving photon-counting detector using polychromatic X-ray attenuation by correcting for the beam hardening effect and detector response. Appl Radiat Isot. 2021;170:109617. https://doi.org/10.1016/j.apradiso.2021.109617.
16. Kimoto N, et al. A novel algorithm for extracting soft-tissue and bone images measured using a photon-counting type X-ray imaging detector with the help of effective atomic number analysis. Appl Radiat Isot. 2021;176:109822. https://doi.org/10.1016/j.apradiso.2021.109822.

17. Reza S, et al. Semiconductor radiation detectors: technology and applications. New York: CRC Press; 2017. p. 85–108.
18. Hayashi H, et al. Advances in medicine and biology. New York: Nova Science Publishers, Inc.; 2019. p. 1–46.
19. Hayashi H, et al. Photon counting detectors for X-ray imaging: physics and applications. Cham: Springer; 2021. p. 1–119. https://doi.org/10.1007/978-3-030-62680-8.
20. Samei E, Flynn JM. An experimental comparison of detector performance for direct and indirect digital radiography systems. Med Phys. 2003;30(4):608–22. https://doi.org/10.1118/1.1561285.
21. Spekowius G, Wendler T. Advances in healthcare technology: shaping the future of medical care. Dordrecht: Springer; 2006. p. 49–64.
22. Knoll GF. Radiation detection and measurement. New York: Wiley; 2000. p. 1–802.
23. Heismann BJ, et al. Density and atomic number measurements with spectral x-ray attenuation method. J Appl Phys. 2003;94(3):2073–9. https://doi.org/10.1063/1.1586963.
24. Hubbell JH. Photon mass attenuation and energy-absorption coefficients. Int J Appl Radiat Isot. 1982;33(11):1269–90. https://doi.org/10.1016/0020-708X(82)90248-4.
25. Brooks RA, Di Chiro G. Beam hardening in x-ray reconstructive tomography. Phys Med Biol. 1976;21(3):390–8. https://doi.org/10.1088/0031-9155/21/3/004.
26. Good MM, et al. Accuracies of the synthesized monochromatic CT numbers and effective atomic numbers obtained with a rapid kVp switching dual energy CT scanner. Med Phys. 2011;38(4):2222–32. https://doi.org/10.1118/1.3567509.
27. Johnson TRC. Dual-energy CT: general principles. Am J Roentgenol. 2012;199(5):S3–8. https://doi.org/10.2214/AJR.12.9116.
28. Tatsugami F, et al. Measurement of electron density and effective atomic number by dual-energy scan using a 320-detector computed tomography scanner with raw data-based analysis: a phantom study. J Comput Assist Tomogr. 2014;38(6):824–7. https://doi.org/10.1097/RCT.0000000000000129.
29. Wang X, et al. Material separation in x-ray CT with energy resolved photon-counting detectors. Med Phys. 2011;38(3):1534–46. https://doi.org/10.1118/1.3553401.
30. Spiers FW. Effective atomic number and energy absorption in tissues. Br J Radiol. 1946;19(218):52–63. https://doi.org/10.1259/0007-1285-19-218-52.
31. Birch R, Marshall M. Computation of bremsstrahlung X-ray spectra and comparison with spectra measured with a Ge(Li) detector. Phys Med Biol. 1979;24(3):505–17. https://doi.org/10.1088/0031-9155/24/3/002.
32. Hsieh SS, et al. Spectral resolution and high-flux capability tradeoffs in CdTe detectors for clinical CT. Med Phys. 2018;45(4):1433–43. https://doi.org/10.1002/mp.12799.
33. Otfinowski P. Spatial resolution and detection efficiency of algorithms for charge sharing compensation in single photon counting hybrid pixel detectors. Nucl Instrum Methods Phys Res A. 2018;882:91–5. https://doi.org/10.1016/j.nima.2017.10.092.
34. Taguchi K, et al. Spatio-energetic cross-talk in photon counting detectors: numerical detector model (PcTK) and workflow for CT image quality assessment. Med Phys. 2018;45(5):1985–98. https://doi.org/10.1002/mp.12863.
35. Trueb P, et al. Assessment of the spectral performance of hybrid photon counting x-ray detectors. Med Phys. 2017;44(9):e207–14. https://doi.org/10.1002/mp.12323.
36. Zambon P, et al. Spectral response characterization of CdTe sensors of different pixel size with the IBEX ASIC. Nucl Instrum Methods Phys Res A. 2018;892:106–13. https://doi.org/10.1016/j.nima.2018.03.006.
37. Hirayama H, et al. The EGS5 code system. KEK Report. 2005-8 SLAC-R-730. 2005. p. 1–418.
38. Sasaki M, et al. A novel mammographic fusion imaging technique: the first results of tumor tissues detection from resected breast tissues using energy-resolved photon counting detector. Proc SPIE. 2019;10948:1094864. https://doi.org/10.1117/12.2512271.
39. Kuhlman JE, et al. Dual-energy subtraction chest radiography: what to look for beyond calcified nodules. Radiographics. 2006;26(1):79–92. https://doi.org/10.1148/rg.261055034.

40. MacMahon H, et al. Dual energy subtraction and temporal subtraction chest radiography. J Thorac Imaging. 2008;23(2):77–85. https://doi.org/10.1097/RTI.0b013e318173dd38.
41. McAdams HP, et al. Recent advances in chest radiography. Radiology. 2006;241(3):663–83. https://doi.org/10.1148/radiol.2413051535.
42. Kappadath SC, Shaw CC. Quantitative evaluation of dual-energy digital mammography for calcification imaging. Phys Med Biol. 2004;49(12):2563–76. https://doi.org/10.1088/0031-9155/49/12/007.
43. Blake MG, Fogelman I. Technical principles of dual energy X-ray absorptiometry. Semin Nucl Med. 1997;27(3):210–28. https://doi.org/10.1016/S0001-2998(97)80025-6.
44. Cullum DI, et al. X-ray dual-photon absorptiometry: a new method for the measurement of bone density. Br J Radiol. 1989;62(739):587–92. https://doi.org/10.1259/0007-1285-62-739-587.
45. Theodorou JD, et al. Dual-energy X-ray absorptiometry in diagnosis of osteoporosis: basic principles, indications, and scan interpretation. Compr Ther. 2002;28(3):190–200. https://doi.org/10.1007/s12019-002-0028-6.
46. Angelopoulos C, et al. Digital panoramic radiography: an overview. Semin Orthod. 2004;10(3):194–203. https://doi.org/10.1053/j.sodo.2004.05.003.
47. Izzetti R, et al. Basic knowledge and new advances in panoramic radiography imaging techniques: a narrative review on what dentists and radiologists should know. Appl Sci. 2021;11(17):7858. https://doi.org/10.3390/app11177858.
48. Horner K, Devlin H. Clinical bone densitometric study of mandibular atrophy using dental panoramic tomography. J Dent. 1992;20(1):33–7. https://doi.org/10.1016/0300-5712(92)90007-y.
49. Langlais R, et al. The cadmium telluride photon counting sensor in panoramic radiology: gray value separation and its potential application for bone density evaluation. Oral Surg Oral Med Oral Pathol Oral Radiol. 2015;120(5):636–43. https://doi.org/10.1016/j.oooo.2015.07.002.
50. Nackaerts O, et al. Bone density measurements in intra-oral radiographs. Clin Oral Investig. 2007;11(3):225–9. https://doi.org/10.1007/s00784-007-0107-2.
51. Hayashi Y, et al. Assessment of bone mass by image analysis of metacarpal bone roentgenograms: a quantitative digital image processing (DIP) method. Radiat Med. 1990;8(5):173–8.
52. Matsumoto C, et al. Metacarpal bone mass in normal and osteoporotic Japanese women using computed X-ray densitometry. Calcif Tissue Int. 1994;55(5):324–9. https://doi.org/10.1007/BF00299308.
53. Saito M, Sagara S. A simple formulation for deriving effective atomic numbers via electron density calibration from dual-energy CT data in the human body. Med Phys. 2017;44(6):2293–303. https://doi.org/10.1002/mp.12176.

K-Edge Imaging in Spectral Photon-Counting Computed Tomography: A Benchtop System Study

Chelsea A. S. Dunning, Devon Richtsmeier, Pierre-Antoine Rodesch, Kris Iniewski, and Magdalena Bazalova-Carter

1 Computed Tomography Evolution

1.1 Conventional CT

Computed Tomography (CT) has become a mainstay in diagnostic imaging, and 80 million CT scan examinations are performed per year in the USA alone. Since its emergence in the 1970s, conventional CT scanners use scintillator-based energy integrated detectors (EIDs) that, as the name implies, integrate photon energy information. Conventional CT has undergone many design evolutions including multi-slice CT, helical trajectories, and dual-source geometry for fast examinations. However, EIDs have the following limitations which lead to numerous problems in image reconstruction and analysis. First, the X-ray beam is susceptible to beam hardening effects, so higher-energy photons contribute more to the overall signal. Second, CT numbers, expressed in Hounsfield units, differ between systems, making quantitative CT difficult. Third, the contrast-to-noise ratio between two tissues in the body is energy dependent, and low-energy photons usually carry more information. Finally, electronic noise is integrated into the signal, which adversely impacts image quality in CT screening examinations. Despite these shortcomings, the high clinical impact of CT has led to rapidly increased utilization, though some public concern has also been raised about the potential radiation risk to patients.

C. A. S. Dunning
University of Victoria, UVic, Victoria, BC, Canada

Department of Radiology, Mayo Clinic, Rochester, MN, USA

D. Richtsmeier · P. -A. Rodesch · M. Bazalova-Carter
University of Victoria, UVic, Victoria, BC, Canada

K. Iniewski (✉)
Redlen Technologies, Saanichton, BC, Canada

© The Author(s), under exclusive license to Springer Nature Switzerland AG 2023
S. Hsieh, K. (Kris) Iniewski (eds.), *Photon Counting Computed Tomography*,
https://doi.org/10.1007/978-3-031-26062-9_12

1.2   Dual-Energy CT

To overcome the quantification task limitations, Dual-Energy (DE) CT scanners have been developed, and the first DE protocols were approved by the Federal Drug Agency (FDA) in 2006 [1, 2]. DECT is based on a double x-ray spectra measurement: a low-energy (LE) and a high-energy (HE) spectrum. Several implementations of DECT exist such as rapid switching of the tube voltage between low and high energy, dual x-ray sources operating at different tube voltages, sequentially scanning using low and then high energy, or using dual-layer EID detectors. DECT still uses EID technology and faces the same limitations as conventional CT. However, DECT provides spectral information with the reconstruction of density/atomic number maps, virtual monoenergetic images, non-contrast images, or images with calcium removed. This spectral information has enabled the development of new radiological protocols.

1.3   Spectral Photon-Counting CT

Many technical developments over the year have been introduced to improve performance, increase diagnostic capability, and reduce radiation dose. The key components to even lower radiation dose and better spatial resolution are energy-discriminating photon-counting X-ray detectors which offer smaller pixel pitches, increases in the geometric fill factor, multiple energy windows enabling uniform photon weighting, and high count-rate capability.

Spectral photon-counting computed tomography (SPCCT) is an emerging X-ray imaging technology that addresses the limitations of conventional CT and extends the scope of available diagnostic imaging tools [3]. The main advantage of SPCCT technology is that the energy information from the X-ray spectrum is sampled more accurately, allowing the additional information being produced during photon interactions in the subject material, including K-edge effects, to be more effectively utilized. The K-edge is the sudden increase in X-ray absorption of the atom at the binding energy of its inner electronic shell and is material-specific. As a result, the additional spectral information contributes to better characterization of tissues and materials of interest, indicating the future potential of this X-ray imaging modality.

Most studies on the optimization of spectral and dual-energy imaging systems consider spectral separation, dose, photon flux, and detection efficiency, but mostly focus on applications in CT. In DECT, materials can be differentiated by exposing the tissues to two different X-ray spectra or using a combination detector which collects two different energy ranges. The analysis technique of these dual-energy data provides information related to the varying response of tissues to X-rays of different energies. Thus, material differentiation and elemental decomposition become possible with the application of two different X-ray spectra. A disadvantage of this technique is that materials with similar attenuation curves cannot be distinguished. SPCCT imaging reveals additional quantitative information on the attenuation curves by employing energy-sensitive photon-counting detectors. In addition to the decomposition offered by DECT, SPCCT further allows the spectral decomposition of more than one element with a high atomic number (Z). Different numbers of energy bins, ranging from 2 to 8 [4], have been selected by various companies to build SPCCT scanners.

The first SPCCT scanner was built by Siemens Healthcare in 2012 [5] and was used for research purposes only. It was based on a 1.6-mm-thick CdTe sensor and used two energy thresholds. With a small pixel pitch of 225 μm, this detector was prepared for the high radiation fluxes occurring in clinical CT. Each sub-pixel featured two individually adjustable energy thresholds, enabling contrast optimization and multi-energy scans. The 32-slice hybrid prototype scanner (CounT, Siemens Healthcare) used a CT gantry from a commercially available clinical CT scanner and utilized two independent fan-beam systems with mountings for different CT detectors.

The Siemens development eventually resulted in a commercial release as of September 2021, when the NAEOTOM Alpha system was cleared by the FDA. It is anticipated that all CT manufacturers will bring their SPCCT systems into the market in the coming few years. The new generation of scanners will improve soft-tissue contrast, contrast-to-noise of images, reduce the radiation dose and contrast volumes required for the CT exams, and improve the spatial resolution. We will see improvements in image degradation due to beam hardening artifacts caused by bone and metal implants, which will be eliminated or strongly mitigated.

2 Contrast Agents

2.1 Iodinated Contrast Medium

To overcome the limitations of conventional CT, contrast agents have been developed. They are injected in the human body and their role is to increase the contrast in targeted patient areas for the same irradiation dose [6]. Currently, the only contrast agents approved for CT imaging are based on iodine (intravenous) and barium (oral). Specifically, iodinated contrast media (ICM) are frequently used in CT examinations. However, they require pre- and post-injection acquisitions to characterize the iodine concentration [7]. Additionally, ICM and bone have very similar attenuation properties in a conventional CT image. However, they present a difference in their atomic number and one of the first DECT applications was to differentiate between ICM and bone [2, 4]. It has also been shown that DECT enables an ICM load reduction compared to conventional CT [2, 8]. DECT also suffers from the EID limitations and the ICM/bone differentiation is degraded by the resulting artifacts [4]. For example, this can be critical in the differentiation between calcifications and the contrast-enhanced vessel lumen during coronary artery examinations.

SPCCT takes advantage of PCD technology to improve ICM/bone differentiation [9, 10]. SPCCT enables further ICM load and dose reduction for the clinical tasks involving ICM injection. But despite the CT technology that is used, ICM alone also has identified limitations. Firstly, ICM has a high clearance speed which makes it challenging to perform the acquisitions at its concentration peak in the human body. It has also some known toxicity complications that can be problematic for older patients because kidney function deteriorates with age. Finally, ICM presents an inefficient targetability and a relatively poor sensitivity [11]. No new contrast agents 'for CT imaging' have been approved by the FDA in the last two decades [6, 11]. But at the same time, major breakthroughs have been made in materials research, and especially nanoparticle (NP) development. SPCCT represents the ideal technological leap to facilitate the development and utilization of new contrast agents. It has already been demonstrated that SPCCT is able to perform multi-agent imaging with only one acquisition [8]. Combined with an ICM injection, a different K-edge material can also be injected and imaged in one shot. These materials are presented in the next subsection.

2.2   K-Edge Materials for Clinical Applications

X-ray K-edge imaging exploits the well-known material property known as the K-edge effect. It can be simply illustrated using the example of various attenuation processes: the photoelectric effect (PE), Rayleigh scattering, and Compton scattering (CS). The dominant attenuation process for the energy range of interest (20–160 keV) is the PE, which serves as the basis for K-edge CT imaging. For example, the step change in the absorption coefficient at 80.7 keV for gold is the K-edge effect, as shown in Fig. 1.

The physics of the K-edge effect are well known. A photon that undergoes the PE ejects an electron from its atomic orbital and leaves the atom in an excited state. The ejected electron is known as a "photoelectron" and leaves behind a vacancy in the K-shell orbital it once occupied. The large discontinuity in the total mass attenuation coefficient for gold in Fig. 1 is due to the PE; a photon needs to have an energy that is greater than the binding energy of the orbital shell in order to eject a photoelectron. For example, the K-shell orbital of gold has a binding energy of 80.7 keV. This is the threshold energy an incident photon must have in order to eject a photoelectron from the K-shell orbital of a gold atom. The higher the Z of the atom or the lower the kinetic energy of the photon, the more likely the photon will undergo a photoelectric effect process.

The most commonly used contrast agent, iodine, has a K-edge at 33.2 keV within the range of energies for photon-counting SPCCT detectors which is typically 20–160 keV. However, photons below 40 keV are very heavily attenuated by human body making K-edge detection at 33.2 keV very challenging. The effective K-

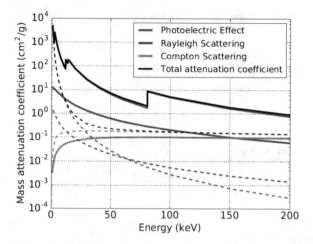

Fig. 1 Total mass attenuation coefficients of gold (solid) and water (dashed), with contributions from each photon interaction type as a function of photon energy. (Data from NIST XCOM tables (found here: https://www.nist.gov/pml/xcom-photon-cross-sections-database))

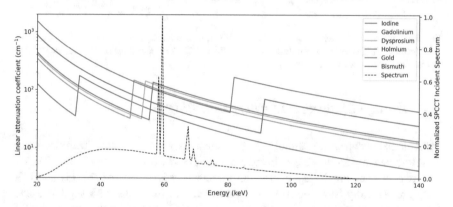

Fig. 2 K-edge contrast agents that have been investigated for SPCCT K-edge imaging

edge range for clinical applications is assumed to be between 40 and 100 keV, and different materials have been identified to correspond to the clinical requirement in terms of toxicity and sensitivity (Fig. 2): gadolinium (50.2 keV), holmium (55.6 keV), ytterbium (61.3 keV), tantalum (67.4 keV), tungsten (69.5 keV), gold (80.7 keV), and bismuth (90.5 keV).

The K-edge effect can be utilized in X-ray imaging to exploit the rapid change of attenuation at the K-edge energy and the fact that it is distinct for every element. In this section, we will focus on multiple novel contrast agents and their combined detection with iodine (I), as they are a mix of typical and novel contrast agents being investigated for medical diagnosis. SPCCT is particularly well suited to detect multiple contrast agents simultaneously as illustrated in Fig. 3.

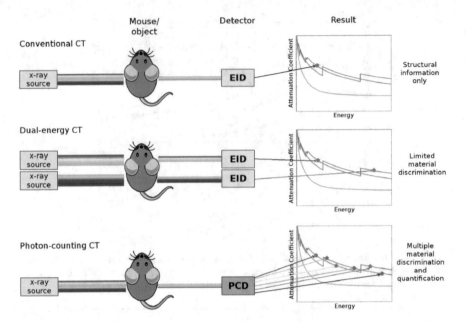

Fig. 3 Diagram of conventional and dual-energy CT enabled by energy integrating detectors (EID) and SPCCT enabled by photon-counting detectors (PCD). (Courtesy of MARS Bioimaging (https://www.marsbioimaging.com/))

2.3 K-Edge Imaging and Energy Binning

Most elements appearing naturally in the human body have low atomic numbers and therefore lack absorption edges in the diagnostic X-ray energy range (20–160 keV). The two dominating X-ray interaction effects in this range are the PE and CS. Both are typically assumed to be smoothly varying functions with separable and independent material and energy dependences. The linear attenuation coefficient μ can therefore be expanded as a linear combination of both processes as:

$$\mu = \alpha_{PE} \times f_{PE} + \alpha_{CS} \times f_{CS} \tag{1}$$

The photoelectric material coefficient (α_{PE}) is assumed to be proportional to mass density and depends strongly on atomic number (Z) (α_{PE} is proportional to product of density ϱ and Z^3), whereas the CS coefficient (α_{CS}) can be described by its density dependence alone (α_{CS} is proportional to the density ϱ). The photoelectric effect (f_{PE}) has a cubic dependence on photon energy E (f_{PE} is inversely proportional to E^3). On the other hand, the energy dependence of the Compton scattering term, f_{CS}, is relatively flat over the diagnostic energy range and is often modeled by the Klein-Nishina cross section.

In contrast-enhanced imaging, one or several high-Z contrast agents with K absorption edges in the diagnostic energy range may be present in the body. K-edge energies are material-specific, which means that the energy dependence of the photoelectric effect is no longer separable from the material properties, and an additional term (or terms if multiple materials with K-edges are present) can be added to Eq. (1) to describe the additional attenuation.

Equation (1) gives the linear attenuation coefficient at a single energy but can be used to calculate the linear attenuation coefficient over an x-ray spectrum via simple sum of coefficients calculated at each of the spectrum energies. This yields a conventional non-energy-resolved image, but because X-ray contrast varies with energy, a correspondingly weighted sum is optimal from a contrast-to-noise ratio aspect. The benefit of so-called energy weighting is most prominent in energy regions where the photoelectric effect is non-negligible, at low energies and/or with high-Z materials because of its strong energy dependence, and optimal weights approximately follow the energy dependence of the photoelectric effect (E^{-3}). For example, in the case of breast tissue, the intersection between the photoelectric and Compton cross sections occurs below 30 keV, above which the efficiency of energy weighting is limited. This implies that the low-energy detection range should be pushed below 30 keV as much as possible by the system implementation, though it will be limited by the level of electronic noise present.

Besides energy weighting, a second approach to make use of the energy information is to treat the acquisitions according to Eq. (1) as a system of equations with material thicknesses as unknowns, a technique broadly referred to as material decomposition. For imaging in the absence of a K absorption edge, the limited number of independent functions of energy according to Eq. (1) means that the system of equations can only be solved for two unknowns, and measurements at two energies (here, the number of energy bins (N_E) would be equal to 2) are necessary and sufficient for a unique solution for materials 1 and 2. Materials 1 and 2 are referred to as basis materials and are assumed to make up the object; any other material present in the object will be represented by a linear combination of the two basis materials. The basis material representation can be readily converted to images showing the amounts of photoelectric and Compton interactions, and further to images of effective atomic number and electron density distributions by utilizing the known material dependences of the PE and CS cross sections [12, 13]. While two-material decomposition is a standard procedure in dual-energy CT, it needs to be pointed out that theoretical work on the dimensionality of the linear attenuation coefficient space has resulted in a statistically significant dimension of four, although the last two bases are very weak and might be neglected for most practical purposes.

If a K absorption edge is present in the object of interest, additional unknowns must be added to the system of equations if one or several K absorption edges are present in the imaged energy range. With one K-edge material being present in the object, measurements at three energies ($N_E = 3$) are necessary and sufficient for a unique solution, and two K-edge materials can be differentiated with four energy bins ($N_E = 4$). Three K-edge materials require at least five energy bins to be used ($N_E = 5$).

A traditional method of contrast-enhanced material decomposition is to linearize the system of equations made up by realizations of Eq. (1) by taking the logarithm (separable energy dependences of the system properties and the linear attenuation coefficients are assumed) and solving by Gaussian elimination, a technique commonly referred to as dual-energy subtraction. More advanced methods are common today, for instance, those based on maximum-likelihood algorithms for overdetermined systems.

The performance of SPCCT imaging applications in general increases with larger signal difference between the energy bins. For energy weighting as well as for material decomposition, a large separation in mean energy between the energy bins is generally beneficial because linear attenuation decreases monotonically with energy in the absence of any absorption edges. For medical CT applications, spectral separation is accomplished by using a large peak voltage (kV_{pp}) value for the X-ray tube. Therefore, practical CT implementations might use kV_{pp} values as large as 140 keV or even 160 keV compared to a traditional system using 120 keV. In practice, the spectral separation is limited by the energy resolution of the detector.

The situation is somewhat opposite for K-edge imaging because, in the presence of absorption edges, the attenuation does not decrease monotonically with energy. With the large, singular rise in attenuation at the K-edge energy, the signal in K-edge imaging increases as the measured energies converge on either side of the K-edge energy and decreases as spectral separation increases [14]. Photon-counting detectors are most suitable for K-edge imaging because the energy bins can be set arbitrarily to coincide with the K-edge energy.

3 K-Edge Imaging Studies

3.1 I, Gd, and Ho K-Edge Imaging

To investigate the K-edge image quality of various contrast agents, a 3D-printed phantom containing 6-mm-diameter vials of I, Gd, and Ho solutions at 1% and 5% concentrations was used to study image quality in an X-ray benchtop SPCCT system (Fig. 4). The imaging setup consisted of an x-ray tube, a CZT detector, linear motion stages, a rotation stage, and the phantom (Fig. 4a). The CZT detector (Redlen Technologies, Saanichton, BC, Canada) consisted of two adjacent modules each 8×12 mm^2 in size with a 2-mm-thick, 330-μm-pitch high-flux sensor. The modules were oriented such that the total active area of the detector was 8 mm $\times$ 24 mm^2. The sensor was capable of operating at 250 Mcps/mm^2 rates without any signs of polarization [15]. As specified by the manufacturer, the CZT detector performed a spectral sweep of ^{241}Am to measure the energy resolution. The full width at half maximum (FWHM) of the 59.5 keV gamma ray peak of ^{241}Am was 8.9 keV.

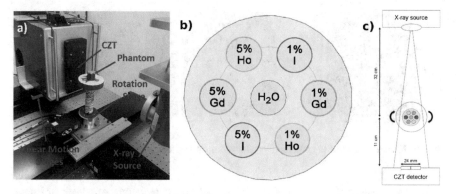

Fig. 4 (**a**) Photo of test setup. (**b**) Schematic diagram of setup (not to scale). (**c**) Top and (**d**) side views of imaging phantom

The energy of photons incident onto the detector was sorted into six energy bins, which were adjustable in the 16 to 200 keV energy range. The MRX-160/22 x-ray tube (Comet Technologies, Flamatt, Switzerland) and the CZT detector were mounted on the M-IMS300V and M-IMS600LM motion stages, and the phantom rested on an RVS80CC rotation stage on an optical table (Newport Corporation, Irvine, CA). The phantom was imaged with a 120 kVp x-ray beam filtered with 1 mm of Al at a beam current of 2 mA and placed at isocenter as shown in the imaging setup (Fig. 4c). Isocenter was located 32 cm from the x-ray source and 11 cm from the detector. The x-ray beam was collimated with lead such that the beam area at the isocenter was 36.6 mm in width and 13.7 mm in height. Each data acquisition was 1 s long, and 180 acquisitions were collected over 360°.

Three of the detector energy bins were tailored to match the K-edges of the three contrast agents (33, 50, and 56 keV). Where possible, the other energy thresholds (16, 63, and 81 keV) were selected to obtain an equal number of counts in each bin above and below the K-edge energies, which was determined based on Monte Carlo simulations of our X-ray source [16].

3.2 3D Image Reconstruction

3D volume datasets with $120 \times 120 \times 24$ voxels of 330 μm in size were reconstructed using the Feldkamp-Davis-Kress (FDK) cone beam algorithm [17] paired with a ring artifact removal filtering algorithm for each of the six energy bins and for the integrated image in MATLAB. A weighted K-edge subtraction for visualization of the three contrast agents was applied as described in Dunning et al. [16], in which K-edge images were equal to the sum of a weighted linear combination of the CT images reconstructed at each of the six energy bins. The

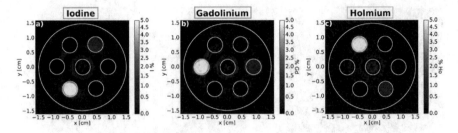

Fig. 5 K-edge images of (**a**) iodine, (**b**) gadolinium, and (**c**) holmium

contrast-to-noise ratio (CNR) of the different contrast vials as a surrogate of image quality was calculated in the following manner:

$$CNR = \frac{\mu_{vial} - \mu_{water}}{\sigma_{phantom}} \qquad (2)$$

where μ and σ are the mean and standard deviation of the contrast vial, water vial, and phantom body regions, respectively.

3.3 Image Quality Results

Figure 5 shows the K-edge reconstructed images for each of the three contrast agents. Both the 5% and 1% concentrations of each of the contrast agents are visible, though there is some noise due to ring artifacts in the images. Figure 6 demonstrates that in K-edge images, the actual concentration at each point in the image can be determined with good accuracy (less than 0.2% difference).

Image quality depends on the spectral distortions introduced by the detector. The main two effects causing spectral distortions are pileup and charge-sharing [3]. To avoid pileup, all CT acquisitions have been conducted under moderate flux conditions preventing flux to be higher than 50 Mcps/mm^2 which according to the detector manufacturer makes pileup distortions smaller than 5%. To study the impact of charge-sharing, two detector modes were used to acquire the data: all raw counts (RAW), or charge-sharing discrimination (CSD) mode in which all detected charge-shared counts are rejected and only the count with the highest energy is recorded. The effect of RAW vs. CSD on I, Gd, and Ho K-edge images is shown in Fig. 7. The RAW I K-edge image has a higher CNR compared to the CSD I K-edge image. This indicates that the CSD algorithm may overcompensate the charge-sharing effect at the two lowest energy bins (16–33 keV and 33–50 keV). However, in the Gd and Ho K-edge images, the CNR is higher using CSD compared to RAW. Above 50 keV, the CSD algorithm appears to benefit the K-edge image quality. Clearly, more studies on the charge-sharing compensation effect on image quality are required.

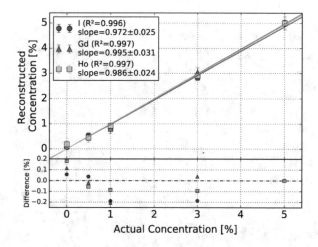

Fig. 6 Reconstructed vs. actual concentration of each contrast agent. The slope and R^2 value are shown in the legend. The bottom subplot shows the percent difference of the reconstructed and actual concentration

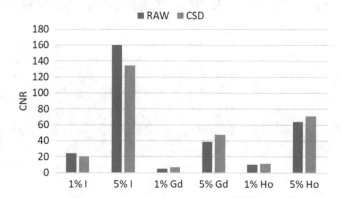

Fig. 7 CNR of each vial in I, Gd, and Ho K-edge images using RAW and CSD data

4 K-Edge Imaging Optimization for Multi-contrast Imaging

Though it has been demonstrated that SPCCT is able to image multiple contrast agents, each SPCCT system must be optimized for the best results depending on the desired image quality and contrast agent(s) used. In this work [18], we demonstrated the experimental variation of a few parameters on a benchtop SPCCT system to obtain the best results as measured by the resulting K-edge contrast-to-noise ratio (CNR). The adjusted parameters discussed were filter type and thickness and energy bin width due to their effect on image quality in previous research [19–21]. A set of practices was developed to choose the correct parameters in the SPCCT system based on the contrast agent.

a)

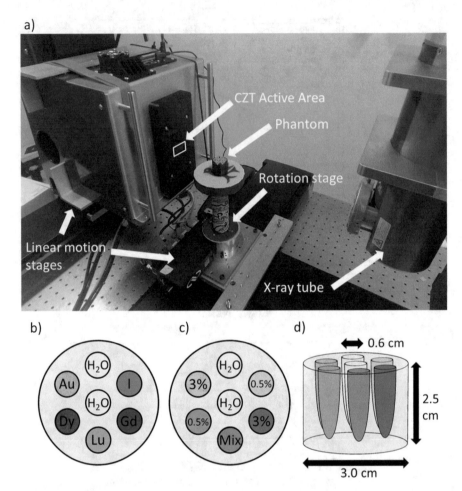

Fig. 8 (**a**) Experimental setup. (**b**) The lanthanide phantom layout. (**c**) The AuGD phantom layout. (**d**) Phantom dimensions

4.1 K-Edge Imaging Optimization Setup and Parameters

The imaging setup and phantoms used in this study are shown in Fig. 8 and were imaged on the same benchtop system and detector as in Sect. 3.1. The phantom (Fig. 8b, c) contained solutions of gold, gadolinium, dysprosium, or lutetium of various concentrations. Two phantom configurations were used in the study, the first of which is shown in Fig. 8b, (called the lanthanide phantom), with a concentration of 3% by weight for all materials. The second phantom, hereby referred to as the AuGd phantom, is shown in Fig. 8c, with gold and gadolinium concentrations of 3% and 0.5% and with a mixture vial consisting of 0.5% concentrations of both materials.

Filter type and thickness and bin width were examined with different phantoms and setup parameters. All acquisitions were captured with a tube voltage of 120 kVp, while the tube current varied between different setups and is detailed below. Three different filter setups were investigated for this study: 2.0 mm Al, 0.5 mm Cu, and 1.0 mm Cu. Tube currents of 1.0 mA, 2.25 mA, and 4.75 mA were used for 2.0 mm Al, 0.5 mm Cu, and 1.0 mm Cu, respectively, in order to keep the number of counts the same in each acquisition. The lanthanide phantom was scanned with bin thresholds set to 16, 50, 54, 64, 81, and 120 keV to accommodate the K-edges of gadolinium (50.2 keV), dysprosium (53.8 keV), lutetium (63.3 keV), and gold (80.7 keV).

For the optimization of bin width, the AuGd phantom was imaged using 0.5 mm Cu filtration and a tube current of 2.25 mA. The energy thresholds were shifted to create different bin widths on either side of the K-edges of the two contrast materials. The two energy thresholds that remained the same were those set at 50 keV and 81 keV, corresponding to the K-edges of gadolinium and gold. Four different scans were taken from the AuGd phantom corresponding to bin widths of 5, 8, 10, and 14 keV for gadolinium and bin widths of 5, 10, 14, and 20 keV for gold. The CT images from each of the corrected projection datasets were reconstructed using the FDK algorithm with a Hamming filter [17].

4.2   Optimization Results

Figure 9 depicts K-edge images using the three filters for dysprosium (a–c) and gold (d–f), both at 3% concentration. For dysprosium, an increase in noise is seen for harder filtrations. For gold the opposite is true, with noise decreasing with softer filtration. Table 1 shows the K-edge CNR for all four contrast agents with each filter type.

As seen in Fig. 9, image noise increased with harder filtration for dysprosium and decreased with harder filtration for gold. This noise trend can also be seen in Table 1, where the K-edge CNR increased with harder filtration for contrast agents with atomic numbers above 71, while CNR decreased for contrast agents with Z-values below 71. This is an effect of the total number of x-ray counts being maintained over the different filters, and as a result, the counts in each of the different energy bins between the spectra of each filtration. A constant image noise level in conventional CT images is desirable because the anatomical information gained from them is necessary to localize the contrast agents. With this restriction, filter choice makes a difference on the detectability of different contrast agents based on their effective atomic number due to the relative number of counts in the two bins used to construct the K-edge images of each contrast agent. Therefore, for contrast agents with Z-values below 71, 2.0 mm Al offers the best results of the three filters tested. For those at or above 71, 0.5 mm Cu gives the highest K-edge CNR.

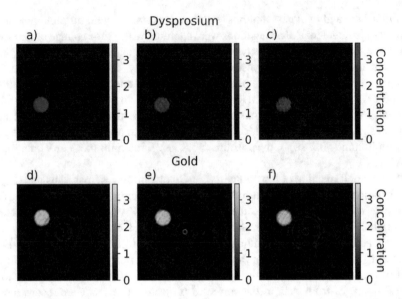

Fig. 9 (**a–c**) K-edge image of 3% concentration dysprosium with (**a**) 2.0 mm Al, (**b**) 0.5 mm Cu, and (**c**) 1.0 mm Cu. (**d–f**) K-edge image of 3% concentration gold with (**d**) 2.0 mm Al, (**e**) 0.5 mm Cu, and (**f**) 1.0 mm Cu

Table 1 K-edge CNR by contrast agent (3% concentration) and filter type

Filter type	Contrast agent CNR			
	Gd ($Z = 64$)	Dy ($Z = 66$)	Lu ($Z = 71$)	Au ($Z = 79$)
2.0 mm Al	3.5 ± 1.4	10.6 ± 1.3	19.9 ± 1.3	12.6 ± 1.3
0.5 mm Cu	8.8 ± 1.4	7.9 ± 1.3	19.0 ± 1.4	15.8 ± 1.3
1.0 mm Cu	5.6 ± 1.5	4.2 ± 1.4	13.5 ± 1.4	14.9 ± 1.4

Figure 10a, b demonstrates how bin width affects the CNR of gadolinium and gold K-edge images, respectively. The data was fit with a quadratic curve to show how the K-edge CNR would likely behave over the range of bin width data that was collected. The ideal bin width must be found for every contrast material separately, as the bin width that results in the highest K-edge CNR can vary, as demonstrated. For gadolinium, peak K-edge CNR was achieved with a bin width of 10.7 keV, and for gold, the optimal bin width was 15.8 keV. The resulting K-edge images reconstructed using the data closest to these peaks (10 keV for gadolinium and 14 keV for gold) are shown in Fig. 10c, d.

The energy bin selection shows the necessity of setting the ideal bin widths in order to maximize CNR to best separate and localize contrast agents. Meng et al. [14] stated that to get the best contrast resolution, the signal-to-noise ratio (SNR) must be maximized. The first contributing factor to SNR in K-edge images is the signal, which is determined by difference in the average linear attenuation coefficient of the contrast agent in the bin above the K-edge and the bin below it.

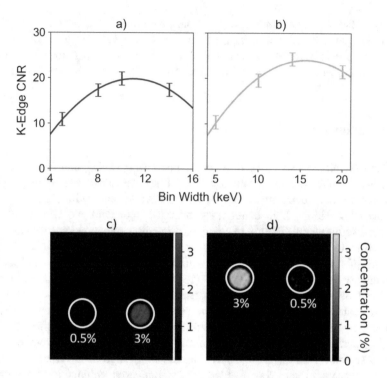

Fig. 10 CNR vs. energy bin width plots for (**a**) gadolinium and (**b**) gold. K-edge images of (**c**) gadolinium and (**d**) gold reconstructed with the highest CNR bin width

The second factor affecting K-edge SNR is, of course, noise. This is determined by the number of x-rays falling in each of the two bins on either side of the K-edge, which follows Poisson statistics. So, if photon flux is constant, adjusting the bin width results in a different number of photons that would fall in that energy range. Ideally, in terms of the signal, the smallest bin size possible would be best to get the largest difference in linear attenuation and thus the most signal in the K-edge image. However, this results in a very small number of x-rays, increasing the noise significantly. Thus, a balance between the two must be found to maximize the SNR.

The differences in the bin width that maximizes gadolinium CNR versus gold CNR can likely be explained by the relative difference in x-ray fluence in the bins around the respective K-edges. The x-ray spectrum filtered with 0.5 mm Cu had more x-rays around the K-edge of gadolinium (50.2 keV) than around the K-edge of gold (80.7 keV). This results in the need to have a wider bin for gold, relative to that for gadolinium in order to increase the counts and reduce the noise to maximize CNR. In addition to reducing noise, the wider optimal bins are also likely a result of the ~8 keV resolution of the detector. Bin widths smaller than 8 keV are likely comprised of fewer counts than what truly falls within that energy range. These optimal bin widths of 10.7 and 15.8 keV are approximately 20% of the K-edge energy of the respective contrast agent. Using 20% of the K-edge energy results in bin width values of 10.0 keV for gadolinium and 16.1 keV for gold. This gives a good rule of thumb for setting the bin widths for these contrast agents, with the 20% values falling within 1 keV of the peak value determined from the quadratic fit.

5   Conclusions

We have simultaneously imaged three contrast agents (I, Gd, and Ho) in a small animal-sized phantom on a tabletop SPCCT system. This was achieved using weighted K-edge subtraction imaging with a CZT detector with six energy bins capable of sustaining high count rates, operated both with and without charge-sharing discrimination. The use of charge-sharing discrimination improved the CNR of the Gd and Ho K-edge images. This first demonstration of multiplexed SPCCT imaging of similar-Z contrast agents, with a difference of Z as low as 3 and K-edge energy as low as 5.4 keV, pushes a technological boundary and could lead to improved disease diagnosis with novel contrast agents.

It has also been demonstrated how parameters in an experimental SPCCT benchtop system can be varied to determine the configuration that offers the best imaging performance for a certain contrast agent. The SPCCT system was also able to separate four different contrast agents in a range of Z-values (64–79) at varying concentrations using K-edge subtraction, even without ideal parameter selection. Possible further improvement of results could be obtained through algorithmic correction of the various mechanisms of energy data distortion, such as charge-sharing. Work on charge-sharing corrections for this detector are currently underway, though the results are mixed, and the algorithm was not applied here. For future use, filter type and thickness will be considered based on Z-value of the contrast agent. The 2.0 mm Al filter will be used for contrast agents with atomic numbers of less than 71, and 0.5 mm Cu used for contrast agents with Z-values of 71 and above. For energy bin width, 20% of the contrast's K-edge energy will be used for gadolinium and gold and verified for other contrast agents before use. This study demonstrates how parameters in a benchtop system can be evaluated in order to obtain the best image quality possible.

References

1. McCollough CH, Leng S, Yu L, Fletcher JG. Dual- and multi-energy CT: principles, technical approaches, and clinical applications. Radiology. 2015;276(3):637. https://doi.org/10.1148/RADIOL.2015142631.
2. McCollough CH, et al. Principles and applications of multienergy CT: report of AAPM Task Group 291. Med Phys. 2020;47(7):e881–912. https://doi.org/10.1002/MP.14157.
3. Taguchi K, Blevis I, Iniewski K. Spectral, photon counting computed tomography: technology and applications. CRC Press; 2020.

4. Willemink MJ, Persson M, Pourmorteza A, Pelc NJ, Fleischmann D. Photon-counting CT: technical principles and clinical prospects. Radiology. 2018;289(2). Radiological Society of North America Inc., 293–312. https://doi.org/10.1148/radiol.2018172656.
5. Kappler S, et al. First results from a hybrid prototype CT scanner for exploring benefits of quantum-counting in clinical CT. SPIE. 2012;8313(2):292–302. https://doi.org/10.1117/12.911295.
6. Yeh BM, et al. Opportunities for new CT contrast agents to maximize the diagnostic potential of emerging spectral CT technologies. Adv Drug Deliv Rev. 2017;113:201–22. https://doi.org/10.1016/j.addr.2016.09.001.
7. Sophie B, et al. CT in non-traumatic acute abdominal emergencies: comparison of unenhanced acquisitions and single-energy iodine mapping for the characterization of bowel wall enhancement. Res Diagn Interv Imaging. 2022;2:100010. https://doi.org/10.1016/J.REDII.2022.100010.
8. Kim J, et al. Assessment of candidate elements for development of spectral photon-counting CT specific contrast agents. Sci Reports. 2018;8(1):1–12. https://doi.org/10.1038/s41598-018-30570-y.
9. Boussel L, et al. Photon counting spectral CT component analysis of coronary artery atherosclerotic plaque samples. Br J Radiol. 2014;87(1040):20130798. https://doi.org/10.1259/BJR.20130798/ASSET/IMAGES/LARGE/BJR.20130798.G006.JPEG.
10. Leng S, et al. Photon-counting detector CT: system design and clinical applications of an emerging technology. Radiographics. 2019;39(3):729–43.
11. Zhang P, et al. Organic nanoplatforms for iodinated contrast media in CT imaging. Molecules. 2021;26(23):7063. https://doi.org/10.3390/MOLECULES26237063.
12. Bazalova M, Carrier JF, Beaulieu L, Verhaegen F. Dual-energy CT-based material extraction for tissue segmentation in Monte Carlo dose calculations. Phys Med Biol. 2008;53(9):2439–56. https://doi.org/10.1088/0031-9155/53/9/015.
13. Bourque AE, Carrier JF, Bouchard H. A stoichiometric calibration method for dual energy computed tomography. Phys Med Biol. 2014;59(8):2059. https://doi.org/10.1088/0031-9155/59/8/2059.
14. Meng B, Cong W, Xi Y, De Man B, Wang G. Energy window optimization for X-ray K-edge tomographic imaging. IEEE Trans Biomed Eng. 2016;63(8):1623–30. https://doi.org/10.1109/TBME.2015.2413816.
15. Iniewski K. CZT sensors for computed tomography: from crystal growth to image quality. J Instrum. 2016;11(12) https://doi.org/10.1088/1748-0221/11/12/C12034.
16. Dunning CAS, et al. Photon-counting computed tomography of lanthanide contrast agents with a high-flux 330-μm-pitch cadmium zinc telluride detector in a table-top system. J Med Imaging. 2020;7(03):1. https://doi.org/10.1117/1.jmi.7.3.033502.
17. Feldkamp LA, Davis LC, Kress JW. Practical cone-beam algorithm. Josa a. 1984;1(6):612–9.
18. Richtsmeier D, Dunning CASS, Iniewski K, Bazalova-Carter M. Multi-contrast K-edge imaging on a bench-top photon-counting CT system: acquisition parameter study. J Instrum. 2020;15(10) https://doi.org/10.1088/1748-0221/15/10/P10029.
19. Roessl E, Proksa R. Optimal energy threshold arrangement in photon-counting spectral x-ray imaging. In: IEEE Nuclear Science Symposium Conference Record, vol. 3, pp. 1950–1954; 2006. https://doi.org/10.1109/NSSMIC.2006.354276.
20. Ren L, et al. The impact of spectral filtration on image quality in micro-CT system. J Appl Clin Med Phys. 2016;17(1):301–15. https://doi.org/10.1120/jacmp.v17i1.5714.
21. Wang AS, Pelc NJ. Optimal energy thresholds and weights for separating materials using photon counting x-ray detectors with energy discriminating capabilities. Med Imaging 2009. 2009;7258(12):725821. https://doi.org/10.1117/12.811454.

Image Domain Performance Evaluation of a Photon-counting ASIC Using a Multilayer Perceptron Network

Brent Budden, Wenhui Qin, Kevin Zimmerman, Yi Qiang, Liang Cai, Zhou Yu, Xiaohui Zhan, Xiaochun Lai, and Richard Thompson

1 Introduction

As an emerging spectral technique, photon-counting computed tomography (PCCT) has great potential to change CT applications. As such, it has garnered enormous research and development interest within both academia and industry. A typical PCCT system employs photon-counting detectors (PCDs) based on room-temperature semiconductor materials such as Cadmium Telluride (CdTe), Cadmium Zinc Telluride (CZT), and Silicon (Si). Unlike conventional CT detectors, PCDs record individual photons' interaction position *and* deposited energy, enabling a CT scanner to reveal a subject's mass attenuation under different x-ray energies, i.e., synthetic mono-energetic scans. This technique has the potential to reduce image noise and radiation dose, improve image resolution, achieve better material decomposition accuracy, enable multiple contrast imaging, and provide for other new clinical applications [8].

An essential task during system development is to compare performance between competing application specific integrated circuit (ASIC) designs. The ASIC is responsible for reading, triggering on, processing, and ultimately digitizing the produced signal from the sensor. Its delicate design has implications on basic detector-level performance metrics such as energy resolution, count-rate response, heat generation, response time and paralysis, thermal sensitivity, and stability. These metrics feed into "downstream" image domain performance metrics including image noise, artifacts, decomposition accuracy, and more.

B. Budden (✉) · K. Zimmerman · Y. Qiang · L. Cai · Z. Yu · X. Zhan · R. Thompson
Canon Medical Research USA, Inc., Vernon Hills, IL, USA
e-mail: bbudden@mru.medical.canon

W. Qin · X. Lai
Biomedical Engineering School, ShanghaiTech University, Shanghai, China

© The Author(s), under exclusive license to Springer Nature Switzerland AG 2023
S. Hsieh, K. (Kris) Iniewski (eds.), *Photon Counting Computed Tomography*,
https://doi.org/10.1007/978-3-031-26062-9_13

One method of comparing performance metrics of any generalized system is to employ simulation techniques complemented by a statistical analysis that involves computing a merit of minimum achievable noise defined by the Cramér-Rao lower bound (CRLB) [2, 7]. Such a technique can be used to derive expected Gaussian-correlated noise for specific systems. While this method can be applied directly to understand noise at the detector level, correlating it to image reconstructions would allow for an image domain analysis which is easier to interpret. Further, while such a technique will *theoretically* produce the desired result, *practically*, it is impeded by high computational cost if employed directly. In this chapter, we present a technique to utilize a multilayer perceptron network to train a model that calculates the noise to inject into simulation-derived sinograms, which are then reconstructed in order to compare ASIC designs in the image domain.

2   Statistical Metric

The Fisher information matrix [3], which is the inverse of the covariance matrix between a number of model parameters of a system, is a means to evaluate that system based on a known model and measurement uncertainties. For N model parameters, the matrix, $\mathcal{F}$, is an $N x N$ matrix. Each element, $\mathcal{F}_{ij}$, of the Fisher information matrix is calculated by

$$\mathcal{F}_{ij} = \frac{\partial \bar{n}^T}{\partial L_i} \Sigma^{-1} \frac{\partial \bar{n}}{\partial L_j} + \frac{1}{2}\text{tr}\left(\Sigma^{-1} \frac{\partial \Sigma}{\partial L_i} \Sigma^{-1} \frac{\partial \Sigma}{\partial L_j} \right), \tag{1}$$

where $\mathcal{F}$ is of size of 2×2 for typical two-basis material (i.e., bone and water) decomposition; L_i is an element of a material length vector L, which has a size of 2×1 for two-basis material decomposition; $\bar{n}$ is expected bin counts; and Σ is a covariance matrix of the bin counts.

CRLB states that the minimum achievable noise of such a system is defined by the inverse of the Fisher information matrix, i.e.,

$$\Sigma_L \geq \frac{1}{\mathcal{F}}. \tag{2}$$

The quantities are typically derived from a simulation by sampling mean and covariance of a large number of random trials. The derivative of both $\bar{n}$ and Σ with respect to a basis material is calculated numerically by repeating the simulation with a small increment of either basis material length. The computational cost of such a process is high when signal response of each photon interaction has to be simulated, as we typically do not have a simple analytical model for sensor and ASIC response. Thus, we can only evaluate a few typical material combinations in the detector domain within a reasonable time frame. Further, a more valuable way would be to assess the decomposition noise or mono-energy image noise in the image space, as this provides a well-accepted criterion which is easier to interpret.

3   Method

Workflow of the complete ASIC performance evaluation is shown in Fig. 1, including the simulation, training, and image domain evaluation. Figure 2 shows the products of the image domain evaluation.

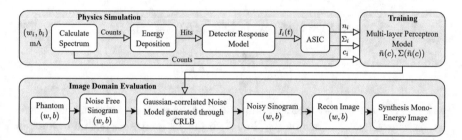

Fig. 1 A flowchart of the complete analysis experiment. A framework comprised of Monte Carlo and analytical simulations informs a multilayer perceptron network, which builds a model of Gaussian-correlated noise. This noise is then fed into an image domain analysis process that ultimately produces synthetic mono-energy images for direct comparison of ASIC implementations in the image domain

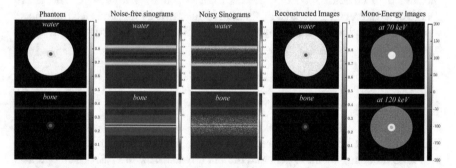

Fig. 2 Example products of the image domain evaluation. An isometric phantom containing both water and bone are used. Noise-free sinograms are forward-projected. Noise was then added according to the trained perceptron. Bone and water images were reconstructed using filtered back projection and mono-energetic images may be synthesized and compared

3.1 Physics Simulation

In this work, we used water and bone basis for material decomposition, i.e., the imaging material is decomposed to the attenuation length of water and bone,

$$L = (w, b). \tag{3}$$

A set of water and bone combinations were simulated. For each combination (w_i, b_i), we generated expected bin counts, n_i, the covariance between bin counts, Σ_i, and rebin counts of the x-ray spectrum attenuated by the water and bone combination, which is given by

$$S_i(E) = S_0(E) \exp\left[-(w_i \mu_w(E) + b_i \mu_b(E))\right] \tag{4}$$

$$c_{ij} = \int_{E_j}^{E_{j+1}} S_i(E) \mathrm{d}E, \tag{5}$$

where $S_0(E)$ is the x-ray tube spectrum and $S_i(E)$ is the x-ray spectrum attenuated by the water/bone combination (w_i, b_i). E_j is the jth energy bin position, which is the same as the ASIC energy threshold position.

To generate the other two outputs (n_i, Σ_i) of the simulation, x-ray interaction with a Cadmium Zinc Telluride (CZT) sensor was modeled through Monte Carlo simulation using GATE [4]. Output of the GATE simulation was hits, which include positions of individual interactions and corresponding deposited energy. The output was fed to a detector modeling package developed by the authors. The initial charge cloud size, charge repulsion, thermal diffusion, trapping, and electronic noise were incorporated in this model by solving charge transportation partial differential equations.[1]

4 ASICs

The output of the detector response modeling, i.e., induced current $I_i(t)$, was fed to the ASIC modeling block to generate the bin counts. How the ASIC then operates on this signal is subject to the application and goals. In the case of PCCT, it is useful to shape the pulse to reduce effects of pileup and to integrate the current in order to derive a voltage pulse whose amplitude is linearly proportional to the total charge (and thereby energy of incident x-ray). This is performed using a charge sensitive amplifier (CSA). In such a configuration, the detector signal output is fed to the inverting input of a CSA such as is seen in Fig. 3 [1]. A feedback capacitor, C_f, and

---

[1] For a detailed analysis of the physics simulation employed in this work, please see [6].

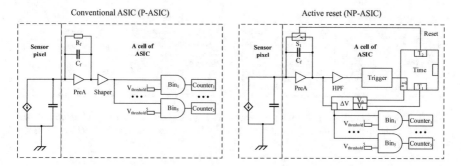

Fig. 3 The two ASIC concepts examined in this work. (Left) A "conventional" ASIC with charge sensitive amplification operated in a paralyzable mode. (Right) An ASIC having an active reset capability operated demonstrating a non-paralyzable mode

a feedback resistor, R_f, are placed in series to the CSA, spanning from the inverting input to the output. The charge from the input signal integrates on the capacitor, and then bleeds off through the feedback resistor, effectively producing a voltage pulse. A system trigger (event above a threshold voltage) causes an active digitization of the pulse height to place the count into one of a number of energy bins.

In a conventional case, the trouble begins in the pileup regime, wherein the incident x-ray flux results in pulses that arrive so close in time such that they significantly overlap. In the best case scenario, the baseline from the first pulse drops back below the threshold before the second pulse arrives. The second pulse may therefore carry incorrect energy information (it is still sitting on the tail of the first pulse), but it is counted. In the worst case scenario, the signal level has not dropped below the threshold voltage for the trigger and thus the second pulse is never counted at all. It is apparent to see that this could continue with additional pulses such that at high enough count rates, the system is "paralyzed"–i.e., does not trigger– and its observed count rate (OCR) versus incident count rate (ICR) would appear to drop back down to zero [5].

One solution to this issue is to employ an active reset scheme. In such an ASIC, a switch is located between the input to the pre-amplifier that effectively grounds the output after a pre-programmable "dead time" has elapsed since the initial trigger. The system may still miss second events that occur within the deadtime of the system, but subsequent events *will* cause a trigger event. In such a case, the system will not be paralyzed ("non-paralyzable"), and its OCR will plateau as a function of ICR.

4.1 Multilayer Perceptron

The output of ASIC was the counts in each energy bin. The expected bin counts $\bar{n}_i$ and the covariance matrix Σ_i were given by the sampled mean and sampled covariance. Two fully connected perceptron networks were structured to capture

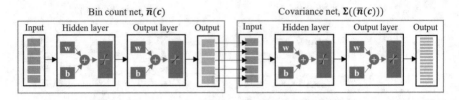

Fig. 4 The two multilayer perceptron networks employed. (Left) A "bin count" network takes rebinned x-ray counts as input to and feeds into (Right) a "covariance" network which generates the noise model

expected bin counts $\bar{n}(c)$ and their respective covariances. Both networks consisted of one input layer, two hidden layers, and one output layer (see Fig. 4). The tanh activation functions were used between input and hidden layers, and linear functions were used between the last hidden layer and the output layer. The first network, referred to as "bin count network," used the rebinned x-ray counts c as input, which is defined in Eq. 5. The output of the bin count network is fed to the second network, which we refer to as "covariance network." Output is the covariance of bin counts. The networks were trained using the simulation data generated in the previous step.

4.2 Image Domain Evaluation

Gaussian random noise was added to the noise-free sinogram of a phantom, which includes both water and bone. The covariance of the Gaussian noise is set to the minimum achievable decomposition covariance, which can be calculated by Eq. 1 and Eq. 2, with $\frac{\partial \bar{n}}{\partial L_i}$ given by

$$\frac{\partial \bar{n}}{\partial L_i} = \frac{\partial \bar{n}}{\partial c} \frac{\partial \bar{c}}{\partial L_i}, \tag{6}$$

and $\frac{\partial \Sigma}{\partial L_i}$ given by

$$\frac{\partial \Sigma}{\partial L_i} = \frac{\partial \Sigma}{\partial \bar{n}} \frac{\partial \bar{n}}{\partial L_i}, \tag{7}$$

where $\frac{\partial \bar{n}}{\partial c}$ and $\frac{\partial \Sigma}{\partial \bar{n}}$ can be calculated by differentiating the bin counts and covariance network, respectively. The noisy sinograms of water and bone were reconstructed using filtered back projection (FBP), and mono-energy images were synthesized using reconstructed water and bone images.

5 Results

For this work, we examined two specific ASIC designs, high-level diagrams of which are shown in Fig. 3. Though the two ASICs we considered differ by more than their detection modes (paralyzable or non-paralyzable), since the detection modes are the primary differentiator, we adopt the names "P-ASIC" and "NP-ASIC" for the conventional and active reset ASICs, respectively.

The method proposed in this study can significantly reduce computing cost. We assume a uniform phantom is scanned by a single slice PCCT with 2700 channels using a 100 mAs protocol. If the image domain study is needed, 18 billion photon events have to be simulated using the conventional way. For the method proposed in this paper, the total events used to train the network are 144 million, i.e., the computing load is reduced by a factor of 1250. It can be done by a workstation with 48 3.0-GHz CPUs within one day.

A multilayer perceptron was implemented according to Sect. 4.1 of this chapter. As output of the training step is the Gaussian-correlated noise model, we may immediately compare expected decomposition noise minima; these results are shown in Fig. 5. The two-dimensional plots show the difference in normalized minimum achievable standard deviation of bone (L1, top) or water (L2, bottom), as a function of the pathlengths L1 and L2, given by:

$$\frac{\delta(L1)}{L1}(\text{NP-ASIC}) - \frac{\delta(L1)}{L1}(\text{P-ASIC}) \text{ and } \frac{\delta(L2)}{L2}(\text{NP-ASIC}) - \frac{\delta(L2)}{L2}(\text{P-ASIC}),$$

(8)

where $\delta(L1)$ and $\delta(L2)$ are minimum achievable standard deviation derived from CRLB, respectively.

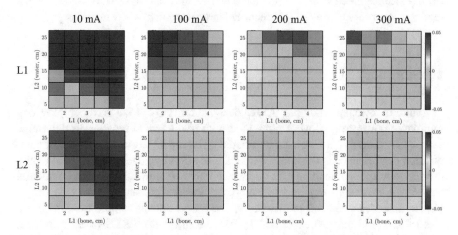

Fig. 5 Relative CRLB of the conventional and non-paralyzable ASICs for each L_b at several tube mAs. (Lower values correspond to where the non-paralyzable ASIC provides superior performance.) Exact equations are shown in Eq. 5

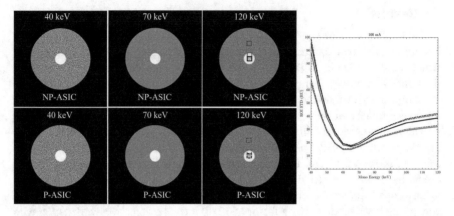

Fig. 6 Synthesized mono-energy images of a mini-soroban phantom, and corresponding HU deviations of 3 ROIs

As is observed for the two specific ASICs evaluated, the NP-ASIC shows better performance for the cases of low flux/hard spectrum, while the P-ASIC shows (minor) better performance for the high flux/soft spectrum cases.

While such analysis is useful for understanding relative performance, it provides no direct human-interpretation of the clinical advantages. Thus, the results are further propagated through the image reconstruction and evaluation domain, as described in Fig. 1, in order to generate products such as are shown in Fig. 2. The results are shown in Fig. 6, where we examine the simulation results of synthetic mono-energy images from a custom-made 10 cm phantom. The phantom contains mostly solid water, with inner inserts of calcium and iodine. Regions are of interest (ROIs) are examined from the reconstructed images the standard deviation of the reported Hounsfield Unit (HU) is recorded as a function of x-ray energies.

6 Summary

In order to evaluate multiple ASIC designs for PCCT, we have employed a multilayer perceptron network combined with traditional simulation to determine the CRLB informing a Gaussian-correlated noise model to evaluate PCCT decomposition performance in the image domain. The effort bypasses the computationally expensive traditional techniques of a large number of individual photon interaction and signal processes required to computed the derivatives necessary to derive CRLB. In order to extend the results to the image domain, where the results are less relative and may be expressed in HU, we inject the CRLB-based minimum achievable noise into simulated sinograms and perform FBP followed by synthetic mono-energetic images. We show that the two ASIC models each provide minor performance enhancements depending on the flux levels and spectral hardness.

The results provide for absolute results here in our study, which are not directly relevant to the reader's use case. However, the intent is to present a novel technique, which the reader may wish to employ for their own work in order to evaluate any component–of course, not limited to the ASIC–to their overall system design. A traditional simulation combined with multilayer perceptron network and Gaussian-based statistical model for minimum achievable noise has the potential to enable a broad range of system designs in a computationally feasible manner.

References

1. Ahmed SN. Physics & engineering of radiation detection. Cambridge: Academic Press, GB; 2007
2. Cramér H. Mathematical methods of statistics. Princeton: Princeton University Press; 1946
3. Fisher RA. On the mathematical foundations of theoretical statistics. Philos Trans R Soc Lond A. 1922;222:309–68
4. Jan S, Santin G, Trul D, et al. GATE: a simulation toolkit for PET and SPECT. Phys Med Biol. 2004;49(19):4543–61
5. Knoll GF. Radiation detection and measurement, 3rd. ed. New York: John Wiley & Sons, Inc.; 2000
6. Lai X, Shirono J, Araki H, et al. Modeling photon counting detector anode street impact on detector energy response. IEEE Trans Radiat Plasma Med Sci. 2020;5(4):476–84
7. Rao CR. Information and the accuracy attainable in the estimation of statistical parameters. Bull Calcutta Math Soc. 1945;37:81–9
8. Willemink MJ, Persson M, Pourmortez A, Pelc NJ, Flesichmann D. Photon-counting CT: technical principles and clinical prospects. Radiology 2018;289(2):293–312

Index